Essentials of Reservoir Engineering

Essentials of Reservoir Engineering

Editor

Urvashi Mehta

Essentials of Reservoir Engineering
Edited by **Urvashi Mehta**

Printed in 2017

ISBN: 978-1-68117-368-9

Library of Congress Control Number: 2015941556

© 2016 by
SCITUS Academics LLC,
616, Corporate Way, Suite 2, 4766,
Valley Cottage, NY 10989

www.scitusacademics.com

Contents

Preface

The primary aim of the book is to present the basic physics of reservoir engineering, using the simplest and most straightforward of mathematical techniques. It is only through having a complete understanding of the physics that the engineer can hope to appreciate and solve complex reservoir engineering problems in a practical manner. This is developed with a homogenous unit system with useful formulas expressed in practical units. Material balance is discussed extensively for oil and gas reservoirs. Attention is given to the importance of the aquifer match before starting a reservoir simulation. Associated gas reservoir development issues are presented. The last chapter is devoted to reservoir simulation. The Theory focuses on the modern tools used in the industry. Calculation techniques are explained to help the user to master the algorithms and optimize the management of the reservoir study. Numerous references are provided to guide the students for further reading. The book content will help students in their first approach to reservoir engineering and professionals to familiarize themselves with modern techniques.

Editor

Metabolic Engineering for Production of Biorenewable Fuels and Chemicals: Contributions of Synthetic Biology

Laura R. Jarboe[1], Xueli Zhang[2], Xuan Wang[2],
Jonathan C. Moore[2], K. T. Shanmugam[2], and
Lonnie O. Ingram[2]

[1]Department of Chemical and Biological Engineering, Iowa State University, 3051 Sweeney Hall, Ames, IA 50011, USA

[2]Department of Microbiology and Cell Science, University of Florida, Gainesville, FL 32611, USA

ABSTRACT

Production of fuels and chemicals through microbial fermentation of plant material is a desirable alternative to petrochemical-based production. Fermentative production of bio renewable fuels and chemicals requires the engineering of biocatalysts that can quickly and efficiently convert sugars to target products at a cost that is competitive with existing petrochemical-based processes. It is also important that biocatalysts be robust to extreme fermentation conditions, biomass-derived inhibitors, and their target products. Traditional metabolic engineering has made great advances in this area, but synthetic biology has contributed and will continue to contribute to this field, particularly with next-generation biofuels. This work reviews the use of metabolic engineering and synthetic biology in biocatalyst engineering for biorenewable fuels and chemicals production, such as ethanol, butanol, acetate, lactate, succinate, alanine, and xylitol. We also examine the existing challenges in this area and discuss strategies for improving biocatalyst tolerance to chemical inhibitors.

INTRODUCTION

Human society has always depended on biomass-derived carbon and energy for nutrition and survival. In recent history, we have also become dependent on petroleum-derived carbon and energy for commodity chemicals and fuels. However, the nonrenewable nature of petroleum stands in stark contrast to the renewable carbon and energy present in biomass, where biomass is essentially a temporary storage unit for atmospheric carbon and sunlight-derived energy. Thus there is increasing demand to develop and implement strategies for production of commodity chemicals and fuels from biomass instead of petroleum. Specifically, in this work we are interested in the microbial fermentation of biomass-derived sugars to commodity fuels and chemicals.

In order for a fermentation process to compete with existing petroleum-based processes, the target chemical must be produced at a high yield, titer and productivity. Sometimes there are additional constraints on the fermentation process, such as the presence of potent inhibitors in biomass hydrolysate or the need to operate at an extreme pH or temperature [1]. These goals can be difficult to attain with naturally-occurring microbes. Therefore, microorganisms with these desired traits often must be developed, either by modification of existing microbes or by the de novo design of new microbes. While significant progress has been made towards de novo design [2, 3], this work focuses on the modification of existing microbes.

Humanity has long relied on microbial biocatalysts for production of fermented food and beverages and eukaryotic biocatalysts for food and textiles. We have slowly modified these biocatalysts by selecting for desirable traits without understanding the underlying biological mechanisms. But upon elucidation of the biological code and the development of recombinant DNA technology, we now have the tools to do more than just select for observable traits—we are now able to rationally modify and design metabolic pathways, proteins, and even whole organisms.

Much of this rational modification has been in the form of Metabolic Engineering. Metabolic Engineering was defined in 1991 [4, 5] and here we use the definition of "the directed improvement of production, formation, or cellular properties through the modification of specific biochemical reactions or the introduction of new ones with the use of recombinant DNA technology" [6]. While Metabolic Engineering has enabled extraordinary advances in the production of commodity chemicals and fuels from biomass, some of which are discussed in this work, we have now reached the point where biological functions that do not exist in nature are desired. Synthetic biology aims to develop and provide these nonnatural biological functions.

For many years, the term Synthetic Biology was used to describe concepts that would be classified today as Metabolic Engineering [7]. However in the last 10 years, terms such as "unnatural organic molecules" [7], "unnatural chemical systems [8], "novel behaviors" [9], "artificial, biology-inspired systems" [10], and "functions that do not exist in nature" [11] have been used to describe Synthetic Biology. For the purpose of this review, we will apply the Synthetic Biology definition of "the design and construction of new biological components, such as enzymes, genetic circuits, and cells, or the redesign of existing biological systems" [12].

Synthetic biology has application to many fields, including cell-free synthesis [13], tissue and plant engineering [14] and drug discovery [15], but here we are interested in the modification of microbes for the biorenewable production of commodity chemicals and fuels. Other recent reviews have also dealt with this topic [16–18].

Synthetic biology for the production of a target compound can be expressed as a sequence of the following events, each of which will be discussed in more detail and demonstrated below. (1) Design the metabolic pathways and phenotypic properties of the desired system. What are the desired substrates and products? What are the expected environmental stressors? (2) Choose an appropriate host organism (chassis) based on the following criteria. Which organisms display at least some of the desired properties? How well characterized and annotated are these organisms? Are there molecular biology tools for modification of this chassis? (4) Formulate an implementation approach. What modifications are necessary to achieve the pathways and properties identified in step (1)? Do metabolic pathways need to be added, removed, or tuned? Does the desired pathway or phenotype exist in nature, or does it need to be designed de novo? (4) Optimize the redesigned system and assess the system properties relative to the ideal. Can the chassis be improved further?

Even a simple biocatalyst, such as the laboratory workhorse Escherichia coli, is a complex system of an estimated 4603 genes, 2077 reactions, and 1039 unique metabolites [19, 20], and while the steps outlined above are relatively straightforward, it is still difficult to quickly and reliably engineer a biocatalyst to perform desired behaviors [21]. Systems biology, the standardization of biological systems, and metabolic evolution are all vital to the compensation for this disconnect between the expected and actual biocatalyst behaviors. Through a combination of these powerful techniques, biocatalysts have been redesigned for the production of an astounding array of commodity fuels and chemicals, both natural and unnatural (Figure 1 and Table 1). Here we discuss successful examples involving the production of commodity fuels and chemicals, with a focus on D- and L-lactate, L-alanine, succinate, ethanol, and butanol.

Table 1: Summary of engineered E. coli biocatalysts for production of renewable fuels and chemicals in our laboratory

Product	Fermentation condition[(1)]	Titer (g/L)	Yield(g/g)	Productivity (g/L/h)	Reference
Redesign through modification of existing pathways					
D-lactate	Anaerobic, batch	118	0.98	2.88	[22]
Acetate	Aerobic, fed-batch	53	0.50	1.38	[23]
Succinate	Anaerobic, batch	83	0.98	0.90	[24]
Redesign through introduction of foreign pathways					
Ethanol	Anaerobic, batch	43	0.48	2.00	[25]
L-lactate	Anaerobic, batch	116	0.98	2.29	[22]
Xylitol	Aerobic, fed-batch	38	1.40	0.81	[26]
L-alanine	Anaerobic, batch	114	0.95	2.38	[27]

[(1)]All fermentations were done in mineral salts medium with glucose, except for the ethanol fermentations which used xylose.

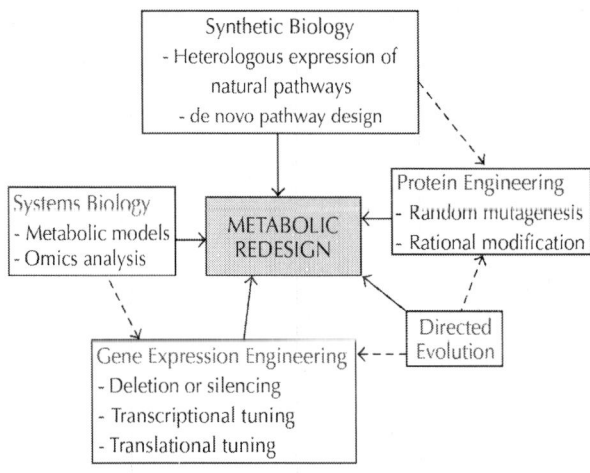

Figure 1: Overview of tools for metabolic redesign.

METHODS AND TOOLS FOR BIOCATALYST REDESIGN

Chassis

A robust and stable chassis enables efficient and economical production of fuels and chemicals at an industrial level. Since we are specifically interested in biocatalysts that can utilize biomass, a desirable chassis has the following characteristics: (1) growth in mineral salts medium with inexpensive carbon sources, (2) utilization of hexose and pentose sugars, so that all the sugar components in lignocellulosic biomass can be converted to the desired product, (3) high metabolic rate, essential for high rate of productivity, (4) simple fermentation process to reduce the manipulation cost and minimize failure risks in large-scale production, (5) robust organism (high temperature and low pH where possible) to reduce the requirement for external cellulase during cellulose degradation, as well as to reduce the required amount of base addition, (6) ease of genetic manipulation and genetic stability, (7) resistance to inhibitors produced during the biomass pretreatment process, and (8) tolerance to high substrate and product concentrations in order to obtain high titers of target compound.

Enteric bacteria, especially E. coli, have many of the above mentioned physiological characteristics and are, thus, an excellent chassis for synthetic biology. Most of the examples discussed here use E. coli, but other important microbial model systems have been redesigned, including Clostridium acetobutylicum [28],Corynebacterium glutamicum [29], Saccharomyces cerevisiae [30], and Aspergillus niger [31]. E. coli has been used as a model organism since the beginning of genetic engineering [32]. While K-12 strain MG1655 (ATCC# 47076) is one of the most commonly used E. coli strains [33], there are other lineages, such as B (ATCC# 11303), C (ATCC# 8739), and W (ATCC# 9637), that are also generally regarded as safe since they are unable to colonize the human gut [34]. Although K-12 is the most characterized and widely used strain, E. coli W (ATCC# 9637) and C (ATCC# 8739) have proven to be better chassis for synthesizing fuels and chemicals. For example, K-12-derived strains were unable to completely ferment 10% (w/v) glucose in either complex or mineral salts medium [1, 35], while derivatives of strains W or C can completely ferment more than 10% (w/v) of glucose with higher cell growth and sugar utilization rates than K-12. Additionally, E. coliW strains have the native ability to ferment sucrose [1, 36].

Foreign genes may be unstable in host cells due to recombination facilitated by mobile DNA elements, and thus the mobile DNA elements in E. coli K-12 strain have been deleted [37]. This minimal genome construction strategy is an excellent approach to improve this chassis for the production of fuels and chemicals.

Systems Biology Tools

Genome-Scale Models and In Silico Simulation

Given the rational basis of metabolic engineering and synthetic biology, models and simulations are critical predictive and tools. Genome sequencing and automatic annotation tools have enabled construction of genome-scale metabolic models of nearly 20 microorganisms [38]. These constraint-based models and in silico simulations can be used to predict metabolic flux redistribution after genetic manipulation, or to predict other cellular functions, such as substrate preference, outcomes of adaptive evolution and shifts in expression profiles [39]. They can also aid in pathway design to obtain desired phenotypes [40–42]. For example, the E. coli iJE660a GSM model was used to successfully simulate single- and multiple-gene knockouts to improve lycopene production [42]. The computational framework, Optknock, was developed to identify gene deletion targets for system optimization [41], and simulation results for gene deletions for succinate, lactate, and

1,3-propanediol production were in agreement with experimental data. Another simulation program, OptStrain, was developed to guide metabolic pathway modification for target compound production, through both the addition of heterologous metabolic reactions and deletion of native reactions [40]. However, most of the current models only have stoichiometric information, while kinetic and regulatory effects are not included [38,39]. Integration of kinetic and regulatory information will improve the accuracy and predictive power of these models.

High-Throughput Omics Analysis

High-throughput omics analysis, such as transcriptome, proteome, metabolome, and fluxome [43–45], aids in characterization of cellular function on multiple levels, and therefore provide a "debugging" capability for system optimization [12, 45].

Genetic manipulations can disturb the metabolic balance or impair cell growth due to depletion of important precursors [46, 47], accumulation of toxic intermediates [48], or redox imbalance [1]. For example, high NADH levels in E. coli reengineered for ethanol production inhibited citrate synthase activity, thereby limiting cell growth by lowering production of the critical metabolite 2-ketoglutarate [49]. Metabolome and fluxome analysis can quickly identify the limiting metabolites or altered metabolic flux distribution, providing the basis for problem solving [45, 50]. For example, metabolite measurements of Aspergillus terreus were implemented in the rational metabolic redesign for increased production of lovastatin [45, 50]. Changes of mRNA and protein profiles can be identified by transcriptome and proteome analysis, providing gene targets for further engineering [46, 47]. The work of Choi et al. demonstrate this concept: transcriptome analysis of E. coli producing the human insulin-like growth factor I fusion protein aided in selection for targets for gene deletion. The resulting redesigned strain showed a greater than 2-fold increase in product titer and volumetric productivity [46, 47]. Additionally, comparative genome sequence analysis facilitates identification of mutated genes or regulators during evolution, and these mutations can be used to redesign the systems for better synthetic capability. For example, in an effort described as "genome-based strain reconstruction", evolved strains of Corneybacterium glutamicum selected for L-lysine production were compared to the parental strain, and mutations were found that were proposed as beneficial to L-lysine production. Three of these mutations were introduced into the parent strain and enabled production of up to 3.0 g/L/hr L-lysine [51].

Genetic Manipulation Tools

Gene Deletion

Gene deletion can redistribute carbon flux toward the target product by deleting genes critical to competing metabolic pathways and, thus, is widely used in metabolic redesign strategies. Homologous recombination is the most frequently used strategy for gene-deletion (Figure 2). Historically, plasmids containing a selectable marker flanked by DNA fragments homologous to the target gene and either temperature sensitive or conditional replicons were needed for efficient gene deletion in bacteria [52] (Figure 2(a)). In contrast, genes can be directly disrupted in yeast by linear PCR fragments with short flanking DNA fragments homologous to chromosomal DNA. Linear DNA is not as easy to transform into E. coli because of the intracellular exonuclease system and low recombination efficiency. Gene deletion systems based on bacteriophage λ Red recombinase facilitate chromosomal gene deletion using a linear PCR fragment [53]. In this method, the chromosomal gene is replaced by the selectable marker flanked by two FRT (FLP recognition target) fragments (Figure 2(b)) and then the marker can be removed by the FLP recombinase [54]. However, this method leaves a 68bp-FRT scar on the chromosome after each excision [52], reducing further gene deletion efficiency. Repeated use of this FRT/FLP system for specific gene deletions has the potential to generate large unintended chromosomal deletions.

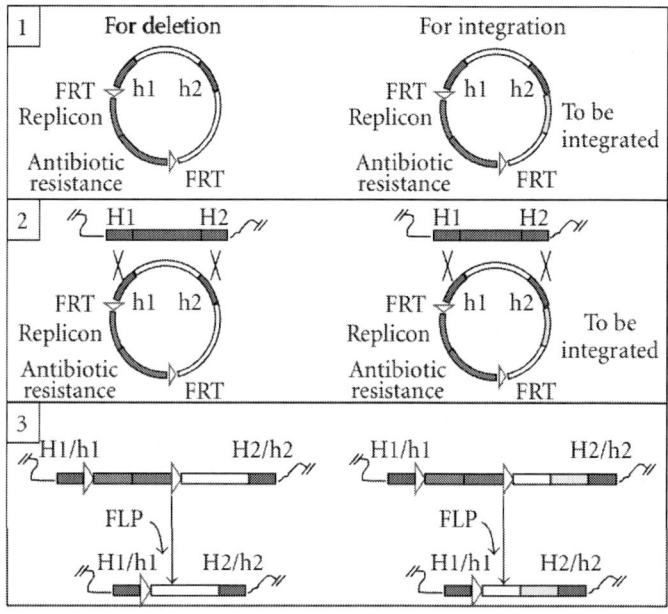

(a)

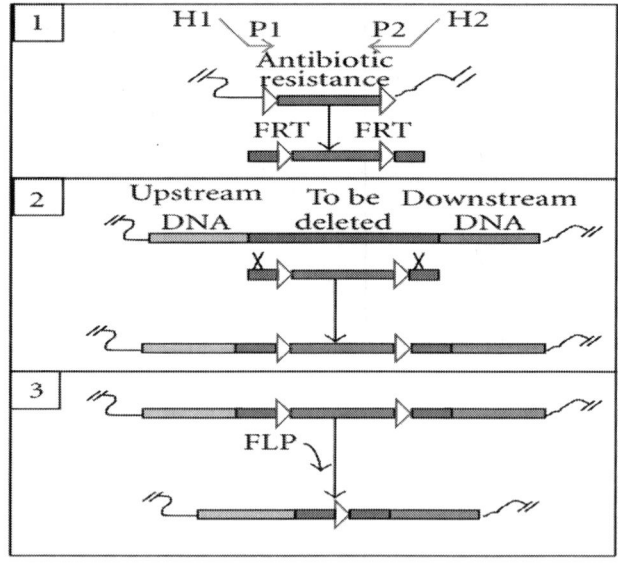

(b)

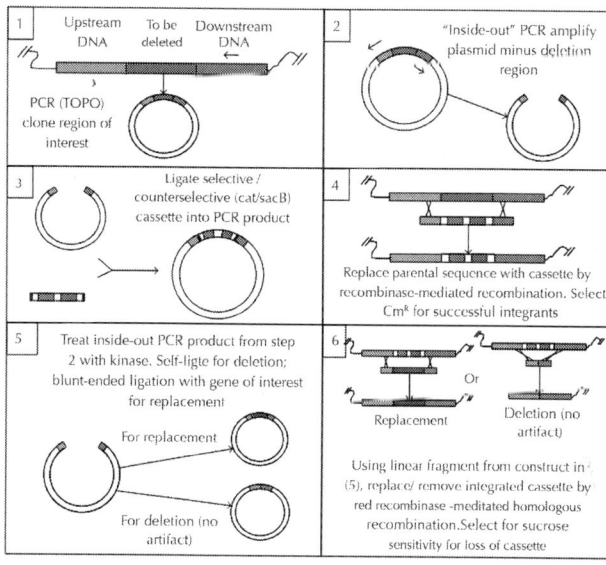

(c)

Figure 2: Comparison of three-gene deletion methods in E. coli. These methods can also be used in other enteric bacteria. The first and third methods can also be used for gene integration into the chromosome and promoter replacement for tuning gene expression. 2(a) plasmid-based method. Step 1 is construction of the deletion plasmid containing DNA fragments homologous to the target gene (h1 and h2), a selectable marker, and either a temperature sensitive or conditional replicon. Step 2 is double-crossover recombination; the plasmid cannot replicate in the host strain, and antibiotic-resistant colonies are selected. In step 3, the FRT, replicon, and antibiotic resistance marker are removed by FLP. 2(b) Linear DNA-based method. Step 1 is construction of the linear DNA fragment by PCR (H1-P1 and H2-P2 as primers). H1 and H2 refer to short DNA fragments homologous to target gene. Step 2 is replacement of the target gene with the antibiotic resistance gene through crossover recombination with the help of Red recombinase. Step is removal of FRT and antibiotic marker by FLP. 2(c) two-stage recombination-based method developed in our lab. Steps 1, 2, 3, and 5 describe construction of the plasmids and linear DNA fragments for the two-stage recombinations. Step 4 describes the first recombination step, in which the cat, sacB cassette is inserted into the target gene. Step 6 is the second recombination step, in which the cat, sacBcassette is removed by selection on sucrose.

To facilitate sequential gene deletions, our lab has developed a two-stage recombination strategy (Figure 2(c)), using the sensitivity of E. coli to sucrose when Bacillus subtilis levansucrase (sacB) is expressed [24, 27, 55]. Gene deletions created by this method do not leave foreign DNA, antibiotic resistance markers, or scar sequences at the site of deletion. In the first recombination, part of the target gene is replaced by a DNA cassette containing a chloramphenicol resistance gene (cat) and levansucrase gene (sacB). In the second recombination, the cat, sacB cassette is removed by selection for resistance to sucrose. Cells containing thesacB gene accumulate levan during incubation with sucrose and are killed [55]. Surviving recombinants are highly enriched for loss of the cat, sacB cassette [24, 27].

Gene Expression Tuning

Like gene deletions, plasmid-based expression systems are ubiquitous to metabolic redesign. However, plasmid-based systems have several disadvantages. (1) Plasmid maintenance is a metabolic burden on the host cell, especially for high-copy number plasmids [56]. Note that high copy numbers are not essential, considering that most central metabolic enzymes are encoded by a single gene; (2) plasmid-based expression is dependent on plasmid stability, with only few natural unit-copy plasmids having the desired stability [12]; (3) only low-copy number plasmids have replication that is timed with the cell cycle, and thus maintaining a consistent copy number in all cells is challenging [12]; (4) metabolic redesign can require construction of a complex heterologous pathway, and thus several genes, encoded in large pieces of DNA, need to be incorporated. Most commercial plasmids have difficulties carrying large DNA fragments.

Chromosomal integration of the target genes followed by fine-tuning their expression could eliminate these plasmid-associated problems. The abovementioned two-step recombination strategy for gene deletion can also be used for gene integration or promoter replacement (Figure 2).

Gene expression in prokaryotes is mainly controlled at the transcriptional level, and therefore the promoter is the most tunable element. While inducible promoters, such as lac and ara, have been traditionally used to modulate gene expression, large-scale inducer use is cost prohibitive for production of fuels and bulk chemicals. However, several strategies have been developed to construct constitutive promoter libraries for fine-tuning gene expression. Some methods rely on the use of natural promoters. For example, Zymomonas mobilis genomic DNA was used to construct a promoter library for screening optimal expression of Erwinia chrysanthemi endoglucanase genes (celY and celZ) in Klebsiella oxytoca P2 in order to improve ethanol production from cellulose [57]. Other methods rely on random modification of existing promoters, such as the randomization of the spacer sequences between the consensus sequences [58], or mutagenesis of a constitutive promoter [59]. This promoter modification method was used to assess the impact of phosphoenolpyruvate carboxylase levels on cell yield and deoxy-xylulose-P synthase levels on lycopene production, and the optimal expression levels of these genes were identified for maximal desired phenotype [59]. These synthetic promoter libraries could also be integrated into the chromosome directly, which could facilitate expression modulation of chromosomal genes [60, 61].

The fine-tuning methods described above rely on the selection of the best natural promoter or random alteration of existing promoters. One of the goals of synthetic biology is construction of standard parts, and posttranscriptional processes, such as transcriptional termination, mRNA degradation, and translation initiation, have been engineered with this goal in mind. Examples include construction of a synthetic library of 5' secondary structures to successfully manipulate mRNA stability [62], and modulation of the ribosome binding site (RBS) as well as Shine-Dalgarno (SD) and AU-rich sequences to tune gene expression at the translation initiation process [60, 63]. Riboregulators were also developed to tune gene expression via RNA-RNA interactions [64]. A final method of fine-tuning gene expression is codon optimization, which can improve translation of foreign genes [65]. These optimized gene sequences often do not exist in nature and must be generated using DNA synthesis techniques.

In many cases, more than one gene needs to be introduced into the chassis and expression of these genes needs to be coordinated to attain desired biocatalyst performance. One such method is modulation of the expression of each individual gene via its own promoter. However, it is difficult to predict the appropriate expression level of each gene. Another option is to combine multiple genes into a synthetic operon with a single promoter, and fine-tune expression of each gene through posttranscriptional processes [12] with tunable control elements (such as mRNA secondary structure, RNase cleavage sites, ribosome binding sites, and sequestering sequences) at intergenic regions. Libraries of tunable intergenic regions (TIGRs) were generated and screened to tune expression of several genes in an operon [48]. This method was used to coordinate expression of three genes in an operon that encodes a heterologous mevalonate biosynthetic pathway, improving mevalonate production by 7-fold [48]. Another method to control expression of more than one gene is to engineer global transcription machinery by random mutagenesis of transcription factors [66, 67]. This method was shown to efficiently improve tolerance to toxic compounds and production of metabolites, and to alter phenotypes [66, 67].

Protein Engineering

Natural proteins may not meet the required criteria for specific and efficient system performance, and thus alteration for a specific application may be needed. Directed evolution of proteins offers a way to rapidly optimize enzymes, even in the absence of structural or mechanistic information [68]. For directed evolution, a protein library is usually generated by random mutagenesis [68], recombination of a target gene [69], or a family of related genes [70] and then the library is analyzed by high-throughput screening. This method has been used to successfully increase enzyme activity [71, 72], increase protein solubility and expression, invert enantioselectivity, and increase stability and activity in unusual environments [68]. For example, a mutation library of the gene-encoding geranylgeranyl diphosphate synthase of Archaeoglobus fulgidus was generated to screen for mutants with higher activity, enabling lycopene production in E. coli. Screening of more than 2,000 variants identified eight with increased activity; one of which increased lycopene production by 100% [71].

Of particular relevance to the field of synthetic biology is the creation of novel enzymatic activity through protein engineering [73, 74]. For example, the unnatural isomerization of -alanine to -alanine was attained by evolving a lysine 2,3-aminomutase to expand its substrate specificity to include -alanine [73].

Rational design is another powerful tool to increase protein properties, especially with the aid of computational analysis [75, 76]. Based on knowledge of protein structure and function, one can predict which amino acid(s) to change in order to obtain the desired function. In the redesign of Lactobacillus brevis for the production of secondary alcohols, it was desired to change the cofactor preference of the R-specific alcohol dehydrogenase from NADPH to NADH. A structure-based computational model was used to identify potentially beneficial amino acid substitutions and one of these changes increased NADH-dependent activity four-fold [77].

While these examples demonstrate the power of rational enzyme (re-)design, this approach requires detailed information about the protein structure and mechanism, while random mutagenesis does not. Recent advances have combined directed evolution and rational design in a so-called "semi-rational" approach to successfully improve enzyme activity when only limited information is available [78, 79]. When the mutagenesis is limited to specific residues, as chosen from existing structural or functional knowledge, these "smart" libraries are more likely to yield positive results [79]. For example, the catalytic activity of pyranose-2-oxidase was improved by mutagenesis of the known active site [80].

While the 20 natural amino acids supply enzymes with a wide range of possible activity, this range can be expanded even further by the use of unnatural amino acids (UAAs). There are more than 40 UAAs available at this time and they have been used to probe protein function, photocage critical residues, and alter metalloprotein properties [81, 82]. While this technology is still in the developmental stage, at least one study has shown an improvement in enzyme activity following insertion of UAAs. Site 124 of E. coli's nitroreductase was replaced with a variety of natural and unnatural amino acids and certain UAA variants had a greater than 2-fold increase in activity over the best natural amino acid variant [83]. This biomimetic approach has been expanded to other metabolites, such as carbohydrates [84] and lipids [85].

Evolution

As described above, a robust biocatalyst with high yield, titer, and productivity is critical for a fermentation process to compete with petrochemical-based production. Current models and simulation tools provide a framework given the constraints of known protein functions. But the many reactions and enzymes that remain uncharacterized cannot be included in this theoretical analysis. Therefore rational design methods often result in a biocatalyst that performs poorly relative to the model. Metabolic evolution provides a complementary approach to improve biocatalyst productivity and robustness, dependent upon the design of an appropriate selection pressure. Where feasible, synthesis of the target compound can be coupled to the production of ATP, redox balance, or key metabolites that are essential for growth, and selection for improvements in growth during metabolic evolution (serial transfers) can be used to coselect for higher rates or titers of target compounds (Figure 3). Both redox balance and net ATP production in such a synthetic system are requisites for successful evolution.

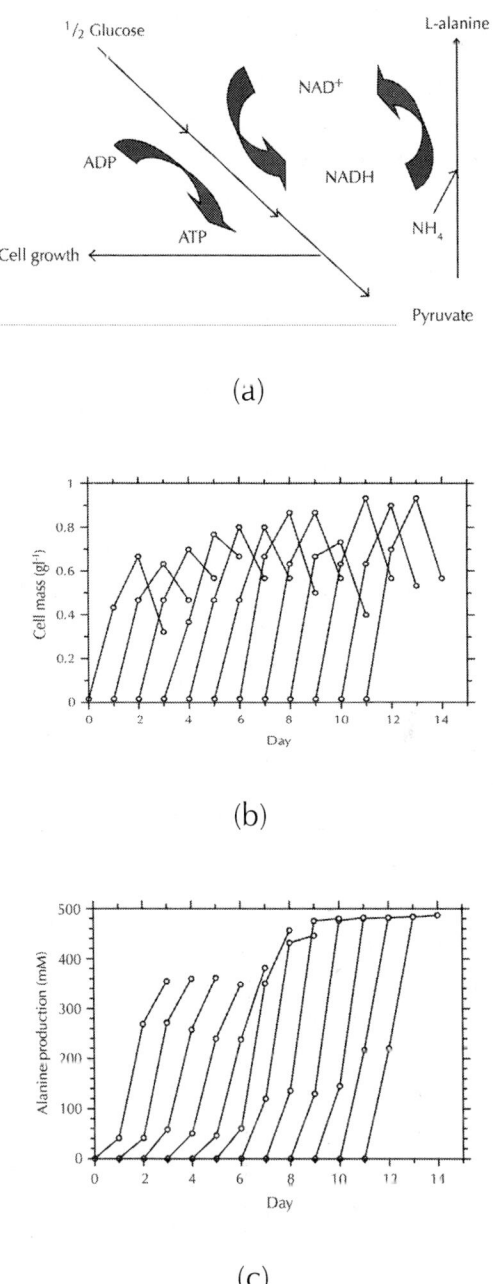

(a)

(b)

(c)

Figure 3: Metabolic evolution for improving L-alanine production in E. coli [27]. 3(a) Redesigned metabolic pathway for L-alanine production:

ATP production and cell growth is coupled to NADH oxidation and L-alanine production. 3(b) Directed evolution improves cell growth. Parental strain XZ112 reaches a maximum cell mass of $0.7\,gL^{-1}$ after 48 hours of fermentation; evolved strain XZ113 attains $0.7\,gL^{-1}$ after 24 hours and a maximum of $0.9\,gL^{-1}$ after 48 hours; 3(c) metabolic evolution to improve cell growth also improves alanine production. Parental strain XZ112 produces 355 mM alanine after 72 hours of fermentation; evolved strain XZ113 produces 484 mM in 48 hours.

We have used this metabolic evolution strategy to optimize biocatalysts redesigned for production of several fermentation products [1], including ethanol, D-lactate, L-lactate, L-alanine (Figure 3), and succinate, as described in more detail below. A frequently-used design scheme is to couple synthesis of the target product to growth by inactivating competing NADH-consuming pathways. Thus, the only way for cells to regenerate NAD^+ for glycolysis is to produce the target compound. Increased cell growth, supported by higher ATP production rate during glycolysis, is coupled with higher NADH oxidization rate, and thus tightly coupled with synthesis of target product. This evolution strategy has been shown to increase productivity by up to two orders of magnitude.

Computational frameworks based on genome-scale metabolic models have been used to construct biocatalysts that couple biomass formation with chemical production [40, 41], and therefore provide a basis for selective pressure for high productivity. For example, Optknock identified gene deletion targets for the construction of lactate-producing E. coli, and then directed evolution improved production capability [86]. Although rational design of metabolic pathways based on current metabolic models is a common method for maximizing yield of the target compound, this method is not always the best strategy, due to our limited understanding of the complicated metabolic network and dynamic kinetics of each reaction. Metabolic evolution provides an excellent alternative method for strain improvement, through which reactions that are not currently predictable would be selected to improve biocatalyst performance [87]. As our knowledge of biocatalyst behavior and metabolism improves, predictive models will become even more powerful.

REDESIGN THROUGH MODIFICATION OF EXISTING PATHWAYS

In this section, we highlight projects that have redesigned a chassis to produce target compounds at high yield and titer without the introduction of foreign pathways. In the next section, we describe biocatalyst redesigns which used foreign or nonnatural pathways.

Succinate

Succinate, a four-carbon dicarboxylic acid, is currently used as a specialty chemical in food, agricultural, and pharmaceutical industries [88] but can also serve as a starting point for the synthesis of commodity chemicals used in plastics and solvents, with a potential global market of $15 billion [89]. Succinate is primarily produced from petroleum and there is considerable interest in the fermentative production of succinate from sugars [89].

Several rumen bacteria can produce succinate from sugars with a high yield and productivity [90–92], but require complex nutrients. Alternatively, native strains of E. coli ferment glucose effectively in simple mineral salts medium but produce succinate only as a minor product [93]. Therefore E. coli strain C (ATCC 8739) was redesigned for succinate production at high yield, titer, and productivity [94].

The initial redesign strategy focused on inactivation of competitive pathways, specifically deletion of lactate dehydrogenase (ldhA), alcohol/aldehyde dehydrogenase (adhE), and acetate kinase (ackA). However, the resulting strain grew poorly in mineral salts medium under anaerobic condition and accumulated only trace amounts of succinate. Because NADH oxidization is coupled to succinate synthesis in this strain, metabolic evolution was used to improve both the cell growth and succinate production. After inactivation of pyruvate formate-lyase and methylglyoxal synthase to eliminate formate and lactate production, the final strain, KJ073, produced near 670 mM succinate (80 g/L) in mineral salts medium with a high yield (1.2 mol/mol glucose) and high productivity (0.82 g/L/h) [94]. Inactivation of threonine decarboxylase (tdcD), 2-ketobutyrate formate-lyase (tdcE), and aspartate aminotransferase (aspC) further increased succinate yield (1.5 mol/mol glucose), titer (700 mM), and productivity (0.9 g/L/h) [24].

Despite its power in improving biocatalyst performance, metabolic evolution has the undesirable property of being a black box; evolved strains show the desired biocatalyst properties, but the metabolic evolution process does not improve our understanding of the biocatalyst. Therefore, reverse engineering of evolved strains can help us identify key mutations that can then be rationally applied to other biocatalysts. Reverse engineering of the succinate-producing strain revealed two significant changes in cellular metabolism that increased energy efficiency [87]. The first change is that PEP carboxykinase (pck), which normally functions in gluconeogenesis during the oxidative metabolism of organic acids [90, 95, 96], became the major carboxylation pathway for succinate production. High-level expression of PCK dominated CO_2 fixation and increased ATP yield (1 ATP per oxaloacetate produced). The second change is that the native phosphoenolpyruvate- (PEP-) dependent phosphotransferase system for glucose uptake was inactivated and replaced by an alternative glucose uptake pathway: GalP permease (galP) and glucokinase (glk). These changes increased the pool of PEP available for maintaining redox balance, as well as increasing energy efficiency by eliminating the need to produce additional PEP from pyruvate, a reaction that requires two ATP equivalents [97].

While rational design based on current metabolic understanding is a key component of metabolic engineering and synthetic biology, our limited understanding of the complicated metabolic network and dynamic kinetics of each reaction can lead to failure of predictive models. In this example, metabolic evolution was demonstrated as an excellent alternative method for strain improvement, through which currently unpredictable reactions would be selected to expand cellular metabolic capability [87]. By understanding the mutations that enabled desirable performance of the succinate-producing strain, we have more options available for the redesign of future systems. To demonstrate this, E. coli was again redesigned based on the findings from the evolved strain [98]. This time, the design strategy shifted from inactivating competitive fermentation pathways to recruiting energy conserving pathways for efficient succinate production (Figure 4(e)). After increasing pck gene expression and inactivating the native glucose PTS system, the native E. colimetabolic system was converted to an efficient succinate synthetic system, equivalent to the native pathway of succinate-producing rumen bacteria [98].

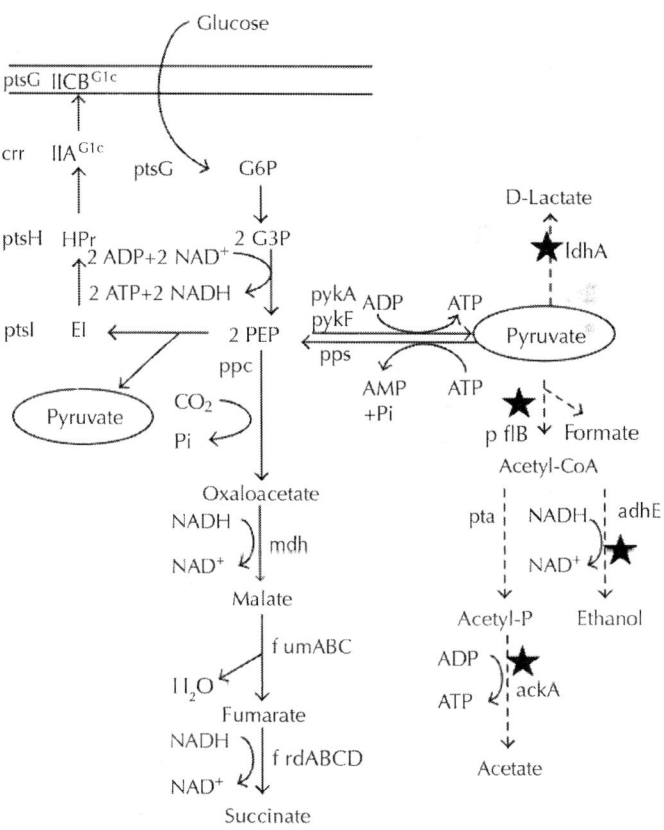

(a)

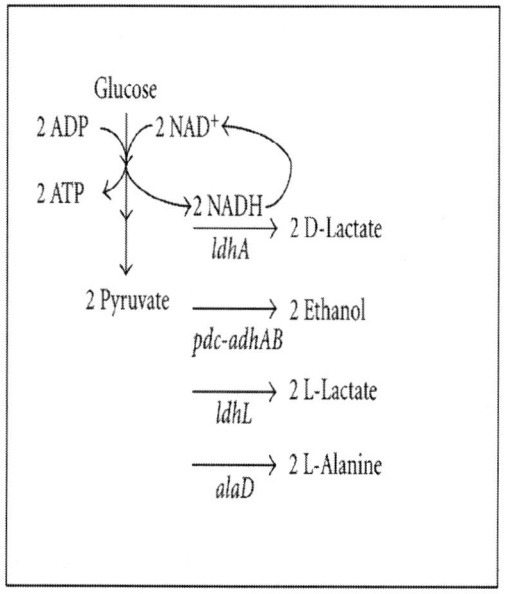

(b)

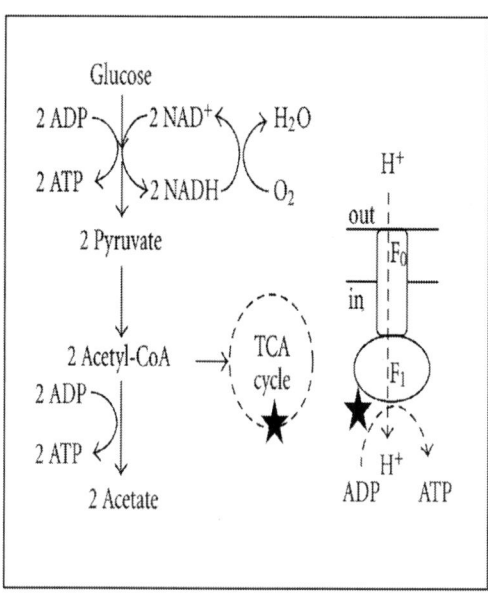

(c)

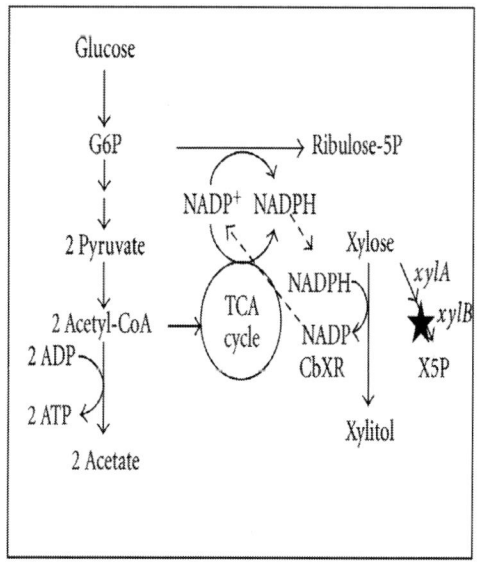

(d)

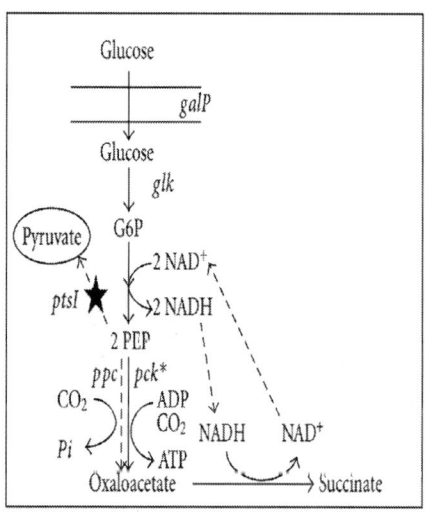

(e)

Figure 4: Synthetic pathways of E. coli for production of fuels and chemicals in our lab: 4(a) Native metabolic pathways of glucose fermentation in E. coli; 4(b) synthetic pathways for production of D-lactate, ethanol, L-lactate and L-alanine; 4(c) synthetic pathways for production of pyruvate and acetate; 4(d) synthetic pathway for production of xylitol, 4(e) synthetic pathway for production of succinate. ★ indicate gene deletion. Genes and enzymes: ackA, acetate kinase; adhAB, alcohol dehydrogenase (Z. mobilis);adhE, alcohol/aldehyde dehydrogenase; alaD, L-alanine dehydrogenase (G. stearothermophilus); crr, glucose-specific phosphotransferase enzyme IIA component; frd, fumarate reductase; fum, fumarase; galP, galactose-proton symporter (glucose permease);glk, glucokinase; ldhA, D-lactate dehydrogenase; ldhL, L-lactate dehydrogenase (P. acidilactici); mdh, malate dehydrogenase; pdc, pyruvate decarboxylase; pflB, pyruvate formate-lyase; ppc, phosphoenolpyruvate carboxylase; pta, phosphate acetyltransferase;ptsG, PTS system glucose-specific EIICB component; ptsH, phosphocarrier protein HPr;ptsI, phosphoenolpyruvate-protein phosphotransferase (Phosphotransferase system, enzyme I); pyk, pyruvate kinase; xrd, xylose reductase (C. boidinii); xylB, xylulokinase. Metabolites: G6P, glucose-6-phosphate; G3P, glycerol-3-phosphate; PEP, phosphoenol pyruvate; X5P, D-xylulose-5-phosphate.

D-Lactate

D-lactate is widely used as a specialty chemical in the food and pharmaceutical industry. It can also be combined with L-lactate for the production of polylactic acid (PLA), an increasingly popular biorenewable and biodegradable plastic [99, 100] whose commercial success obviously depends on the production cost. Although glucose is the current substrate for fermentative production of lactate, it is desirable to produce this commodity chemical from lignocellulosic feedstock, which contains a mixture of sugars. Some lactic acid bacteria have the desirable native ability to produce large amount of D-lactate under low pH condition, where the low pH reduces the process cost [101, 102]. However, these lactic acid bacteria require complex nutrients, and many of them lack the ability to ferment pentose sugars. The lactic acid bacteria that do ferment pentose sugars unfortunately produce a mixture of lactate and acetate and, thus, are not a good chassis for commercial production of D-lactate. While E. coli can ferment many sugars effectively in a simple mineral salts medium, inherent D-lactate productivity is low and other undesirable metabolites are also produced [103]. Therefore, the E. coli metabolic system was redesigned to attain the desired properties of high yield and productivity of D-lactate.

E. coli strain W3110 was used as the chassis for D-lactate production with a redesign strategy that focused on inactivation of competitive fermentation pathways [104]. After deleting the genes encoding fumarate reductase (frdABCD), alcohol/aldehyde dehydrogenase (adhE), pyruvate formate lyase (pflB), and acetate kinase (ackA), the resulting strain, SZ63, can only oxidize NADH via D-lactate synthesis (Figure 4(b)). Although this strain could completely utilize 5% (w/v) glucose in a mineral salts medium with a yield near theoretical maximum (96%), the volumetric D-lactate productivity of 0.42 g/L/h was relatively low compared with lactic acid bacteria [35]. In addition, this strain can neither utilize sucrose nor completely utilize 10% (w/v) sugar [35]. Therefore, an E. coli W derivative strain was chosen as chassis for more robust D-lactate production [35,105]. After redesigning central metabolism so that D-lactate production was the sole means of oxidizing NADH, metabolic evolution was used to further improve cell growth and D-lactate productivity. The resulting strain, SZ194, efficiently consumed 12% (w/v) glucose in mineral salts medium and produced 110 g/L D-lactate [105] with a volumetric productivity of 2.14 g/L/h, a 5-fold increase over the W3110 derivative. The biocatalyst was further optimized by deleting methylglyoxal synthase gene (mgsA) to eliminate L-lactate production, and by metabolic evolution to increase yield and productivity. The final D-lactate producing strain, TG114, could convert 12% (w/v) glucose to 118 g/L D-lactate with an excellent yield (98%) and productivity (2.88 g/L/h) [22].

Acetate

Acetate is a commodity chemical with 2001 worldwide production estimated at 6.8 million metric tons [23]. Biological production of acetate accounts for only 10% of world production, mainly in the form of vinegar, with the remainder of production through petrochemical routes [106–108]. Biological production of commodity chemicals has historically focused on anaerobic production of reduced products, since substrate loss as cell mass and CO_2 is minimal and product yields are high. Contrastingly, acetate is an oxidized chemical, and traditional biological production involves a complex two-stage process: fermentation of sugars to ethanol by Saccharomyces, followed by aerobic oxidation of ethanol to acetate by Acetobacter [106–108]. To enable microbial production of redox-neutral or oxidized products at high yield, the biocatalyst metabolism needs to be redesigned to combine attributes of both fermentative and oxidative metabolisms.

Redesign of E. coli W3110 metabolism for acetate production focused on three major pathways: fermentative metabolism, oxidative metabolism, and energy supply (Figure 4(c)) [23]. The competitive fermentation pathways (pflB, ldhA, frd, adhE) were inactivated to prevent the consumption of common precursor pyruvate, and the oxidative tricarboxylic acid (TCA) cycle was interrupted to reduce the carbon loss as CO_2. Finally, oxidative phosphorylation was disrupted (atpFH) to reduce ATP production while maintaining the ability to oxidize NADH by the electron transport system, thus increasing the glycolytic flux for more ATP production through substrate-level phosphorylation. Although rationally designed, the resulting strain, TC32, had an undesirable auxotrophic requirement for succinate during growth in glucose-minimal medium. Evolution was used to eliminate this auxotrophy and the final strain, TC36, produced 878 mM acetate (53 g/L) in mineral salts medium with 75% of the maximal theoretical yield. Although this is a lower titer than acetate produced from ethanol oxidation by Acetobacter, TC36 has a two-fold higher production rate, requires only mineral salts medium, and can metabolize a wide range of carbon sources in a simple one-step process [23].

Others

Butanol is an excellent alternative transportation fuel with several advantages compared to ethanol, including higher-energy content, lower volatility, less hydroscopicity, and less corrosivity [109]. Redesign of E. coli for butanol production is discussed below. C. acetobutylicum ATCC 824 naturally produces butanol and was redesigned to increase butanol production and decrease coproduct accumulation. Metabolic engineering-type modifications, such as overexpression of the acetone formation pathway to increase formation of butanol precursor butyryl-CoA, inactivation of the transcriptional repressor SolR, and overexpression of alcohol/ aldehyde dehydrogenase all increased butanol production [110– 112]. In an excellent example of synthetic biology-type applications, expression of the butyrate kinase gene was fine-tuned by a rationally designed antisense RNA to increase butanol production [113].

1,2-propanediol (1,2-PD) is a major commodity chemical currently derived from propylene. E. coli naturally produces low amounts of 1,2-PD, and therefore its metabolism was redesigned to produce 1,2-PD at high yield and titer from glucose this was achieved by inactivation of competing pathways (lactate dehydrogenase and glyoxalase I), and overexpression of essential genes of 1,2-PD synthetic pathway (methylglyoxal synthase, glycerol dehydrogenase, and 1,2-PD oxidoreductase) [114]. Evolution was also used in combination with rational design for increased 1,2-PD production [115].

L-valine, an essential hydrophobic and branched-chain amino acid, is used in cosmetics, pharmaceuticals, and animal feed additives [116]. E. coli was redesigned for L-valine production at high yield and titer from glucose through a combination of traditional metabolic engineering and synthetic biology. Traditional metabolic engineering was used to inactivate competing pathways and overexpress acetohydroxy acid synthase I (ilvBN), part of the valine biosynthesis pathway. Unfortunately, the E. coli chassis has regulatory elements that tightly control L-valine biosynthesis, making production of valine at high yield and titer difficult. Feedback inhibition was eliminated by rational site-directed mutagenesis of acetohydroxy acid synthase III. In an excellent demonstration of the gene expression tuning techniques discussed above, transcriptional attenuation of valine biosynthesis genes ilvGMEDA was eliminated by replacing the attenuator leader region with the constitutive tac promoter. Transcriptome analysis and in silico simulation guided selection of additional target genes for amplification and deletion, and the final biocatalyst produced 0.378 g L-valine per g glucose, giving a titer of 7.55 g/L valine from 20%(w/v) glucose [116]. A similar strategy was also used for L-threonine production [117].

REDESIGN THROUGH INTRODUCTION OF FOREIGN OR NONNATURAL PATHWAYS

Foreign Pathways

Ethanol

Ethanol is a renewable transportation fuel. Replacement of gasoline with ethanol would significantly reduce US import oil dependency, increase the national security, and reduce environmental pollution [118]. However, only 9 billion gallons of ethanol were produced in 2008, and all were from corn-based production. Lignocellulose is generally regarded as an excellent source of sugars for conversion into fuel ethanol. It is, thus, desirable to design or obtain biocatalysts that can utilize all the sugar components in lignocellulose and convert them to ethanol with high yield and productivity in mineral salts medium. Native S. cerevisiae and Z. mobilis strains can efficiently convert glucose to ethanol, but cannot utilize pentose sugars. In contrast, E. colistrains can utilize all the sugar components of lignocelluloses but ethanol is only a minor fermentation product, with mixed acids accumulating as the major fermentation product [103]. While recent advances have been made engineering the native E. coli metabolic pathways for ethanol production [119], the most successful example used a foreign metabolic pathway to enable ethanol production from E. coli strain W (ATCC# 9637) [1].

Redesign for ethanol production was decoupled to three parts: construction of a metabolic pathway for production of ethanol as the major fermentation product, elimination of competitive NADH oxidization pathways, and disruption of side-product formation. The Z. mobilis homoethanol pathway (pyruvate decarboxylase and alcohol dehydrogenase) was introduced as a foreign pathway, enabling redox-balanced production of ethanol at high yield [120] (Figure 4(b)). Then fumarate reductase (frd) was disrupted to increase ethanol yield. The resulting strain, KO11, produced ethanol at a yield of 95% in a complex medium [121]. This strain was developed at the dawn of metabolic engineering and has been used to produce ethanol from a variety of lignocellulosic materials, as reviewed in [1].

Although the ethanol production rate of KO11 was as high as yeast, the ethanol tolerance and performance in minimal medium did not meet the desired standards. Therefore strain SZ110, a derivative of KO11 modified for lactate production in mineral salts media [35], was redesigned for ethanol production [122]. As with the design of KO11, redesign of SZ110 was decoupled to construction of an ethanol synthetic pathway, elimination of competitive NADH oxidization pathways, and blockage of side-product formation. However, this redesign strategy also included the acceleration of mixed sugar co-utilization. The lactate producing pathway was disrupted and the Z. mobilis homoethanol pathway was integrated into the chromosome by random insertion to select for optimal expression. The Pseudomonas putida short-chain esterase (estZ) [123] was introduced to decrease ethyl acetate levels in the fermentation broth and decrease the downstream purification cost. In addition, methylglyoxal synthase (mgsA) was inactivated, resulting in co-metabolism of glucose and xylose, and accelerated the metabolism of a 5-sugar mixture (mannose, glucose, arabinose, xylose, and galactose) to ethanol [25]. After using evolution to increase cell growth and production, the final strain, LY168, could concurrently metabolize a complex combination of the five principal sugars present in lignocellulosic biomass with a high yield and productivity in mineral salts medium [25].

L-Lactate

As described above, L-lactate is the major component of the biodegradable plastic PLA. Although many lactic acid bacteria produce L-lactate with high yield and productivity [124], they usually require complex nutrients.E. coli does not have a native pathway for L-lactate production, and therefore introduction of a foreign pathway was necessary.

The strategy for redesigning E. coli W3110 for L-lactate production was to eliminate competitive NADH oxidization pathways and then construct the desired L-lactate synthetic pathway (Figure 4(b)) [125]. The L-lactate production pathway, L-lactate dehydrogenase (ldhL) from Pediococcus acidilactici, was used and its coding region and terminator were integrated into the E. coli chromosome at the ldhA site, so that ldhL could be expressed under the native ldhA promoter. In addition, since the ldhL gene contains a weak ribosomal-binding region, this region was rationally replaced with ldhA's RBS [125]. Following a period of metabolic evolution, the resulting strain, SZ85, synthesized 45 g/L L-lactate in a mineral salts medium with yield near theoretical maximum (94%). However, this strain was a K-12 derivative and displayed the same problems seen with the K12-based D-lactate-producing strain described above, meaning that it was unable to completely ferment high sugar concentrations and had a low productivity (0.65 g/L/h). Therefore, the same design strategy was implemented in an E. coli W (ATCC# 9637) derivative. After further deleting mgsA gene to improve chiral purity and using metabolic evolution to improve cell growth and productivity, the final L-lactate-producing strain, TG108, could convert 12% glucose to 116 g/L L-lactate with an excellent yield (98%) and productivity (2.29 g/L/h) [22].

Xylitol

The pentahydroxy sugar alcohol xylitol is commonly used to replace sucrose in food and as a natural, non-nutritive sweetener that inhibits dental caries [126]. Xylitol can also be used as a building block for synthesizing new polymers [127]. Current xylitol commercial production involves hydrogenation of hemicellulose-derived xylose with an active metal catalyst [127]. Biological-based processes have also recently been developed, but although high xylitol titer was achieved by some yeast, the process requires complex medium with numerous expensive vitamin supplements [128]. While E. coli does not have the native capability to synthesize xylitol, a redesign strategy for strain W3110 was proposed involving a foreign metabolic pathway [26]. In the proposed redesign, glucose would support cell growth and provide reducing equivalents, while xylose would be used as substrate for xylitol synthesis (Figure 4(d)). The design strategy consisted of three major components: enabling co-utilization of glucose and xylose, separation of xylose metabolism from central metabolism, and construction of a xylitol production pathway (Figure 4(d)). In order to enable co-utilization of glucose and xylose, glucose-mediated repression of xylose metabolism was eliminated by replacing the native crp gene with a cAMP-independent mutant (CRP*). Xylose metabolism was separated from central metabolism by deleting the xylulokinase (xylB) gene, preventing the loss of xylose carbon to central metabolism. Finally, xylose reductase and xylitol dehydrogenase from several microorganisms were tested for xylitol synthetic capability, and the NADPH-dependent xylose reductase fromC. boidinii (CbXR) was found to support optimal xylitol production. The final strain, PC09 (CbXR), could produce 250 mM (38 g/L) xylitol in mineral salts medium. The yield was 1.7 mol xylitol per mol glucose consumed, which was improved to 4.7 mol/mol by using resting cells. It was proposed that xylitol production could be further improved by increasing supply of reducing equivalents [129].

L-Alanine

L-alanine can be used with other L-amino acids as a pre- and postoperative nutrition therapy in pharmaceutical and veterinary applications [130]. It is also used as a food additive because of its sweet taste. The annual worldwide production of L-alanine is around 500 tons [131], and this market is currently limited by production costs. The current commercial production process converts aspartate to alanine via aspartate decarboxylase, where aspartate is produced from fumarate by aspartate ammonia-lyase catalysis [27]. An efficient fermentative process with a renewable feedstock such as glucose offers the potential to reduce L-alanine cost and facilitate a broad expansion of the alanine market into other products.

SZ194, a derivative of E. coli W (ATCC# 9637) that was previously engineered for D-lactate production, was used as the chassis for L-alanine production [27] (Figure 4(b)). Alanine production in the native strain uses glutamate- and NADPH-dependent glutamate-pyruvate aminotransferase. It is preferable to produce L-alanine directly from pyruvate and ammonia using an NADH-dependent enzyme, and therefore L-alanine dehydrogenase (alaD) of Geobacillus stearothermophilus was employed. The native ribosome binding site, coding region, and terminator of alaD gene were integrated into the E. coli chromosome at the ldhA site, so that expression of alaD could be controlled by the native promoter of ldhA, a promoter that has worked well for production of D- and L-lactate, as described above. Further redesign focused on elimination of trace amounts of lactate and increasing the L-alanine chiral purity by deleting mgsA and the major alanine racemase gene (dadX). Metabolic evolution increased the final titer and productivity by 15- and 30-fold, respectively (Figure 3). The latest L-alanine producing strain, XZ132, converted 12% glucose to 114 g/L L-alanine with a 95% yield and the excellent volumetric productivity of 2.38 g/L/h [27].

Combining Multiple Foreign Pathways in a Single Chassis

Although the work described above relied on the introduction of a single foreign pathway, there are other excellent examples that employ pathways from more than one organism in a single host.

E. coli was redesigned for 1,3-propanediol production using S. cerevisiae pathway to convert glucose to glycerol and a K. pneumonia pathway to convert glycerol to 1,3-propanediol [132]. E. coli was also redesigned for isopropanol production by combining acetyl CoA acetyltransferase (thl) and acetoacetate decarboxylase (adc) from C. acetobutylicum with the second alcohol dehydrogenase (adh) from C. beijerinckii and E. coli's own acetoacetyl-CoA transferase (atoAD) [133]. Artemisinic acid, a precursor of antimalarial drug artemisin, was produced by E. coli following the combination of a mevalonate pathway fromS. cerevisiae and E. coli, amorphadiene synthase, and a novel cytochrome P450 monooxygenase (CYP71AV1) from Artemisia annua [12, 134].

S. cerevisiae was redesigned for flavanone production by combining Arabidopsis thaliana cinnamate 4-hydroxylase (C4H), Petroselinum crispum 4-coumaroyl: CoA-ligase (4CL), and Petunia chalcone synthase (CHS), Petunia chalcone isomerase (CHI) [135]. A similar synthetic system producing hydroxylated flavonols was also constructed in E. coli with additional amplification of C. roseus P450 flavonoid 3', 5'-hydroxylase (FH) fused with P450 reductase, Malus domestica flavanone 3β-hydroxylase (FHT), and Arabidopsis thaliana flavonol synthase (FLS) [136]. The flavonoid production was significantly increased through further redesigning of the central metabolic system of E. coli to increase precursor (Malonyl-CoA) supply [137].

Modification of Natural Pathways for Production of Unnatural Compounds

One of the goals of synthetic biology is to design or construct new genetic circuits. In the examples given thus far, existing biological parts have been reassembled to engineer a biocatalyst that efficiently produces a product that already exists in nature. However, metabolic pathways can also be constructed to produce unnatural compounds.

As discussed above, directed evolution of proteins can modify their activity such that new substrates are recognized or new products are formed [138]. For example, novel carotenoid compounds were generated by evolution of two key carotenoid synthetic enzymes, phytoene desaturase, and lycopene cyclase [139]. Additionally, combinatorial biosynthesis, which combines genes from different organisms into a heterologous host, can also generate new products [140]. For example, four previously unknown carotenoids were produced by combinatorial biosynthesis in E. coli [141].

De Novo Pathway Design

In order to broaden the available biosynthesis space, it is essential to go beyond the natural pathways and design pathways de novo [142]. Although this exciting design strategy still has many challenges, several successful examples have been reported.

For example, a synthetic pathway for 3-hydroxypropionic acid (3-HP) production was designed involving the unnatural isomerization of -alanine to -alanine, as mentioned above. In this example the researchers used directed evolution to expand the substrate specificity of lysine 2,3-aminomutase to include -alanine [73]. The resulting -alanine can then be converted to 3-HP through existing metabolic pathways.

Unnatural pathways for higher alcohol production in E. coli were designed by combining the native amino acid synthetic pathways with a 2-keto acid decarboxylase from Lactococcus lactis and alcohol dehydrogenase from S. cerevisiae [143]. The 2-keto acid intermediates in amino acid biosynthesis pathways were redirected from amino acid production to alcohol production, enabling production of 3-methyl-1-pentanol. This pathway was then expanded for production of unnatural alcohols by rational redesign of two enzymes, with the resulting biocatalysts having the ability to synthesize various unnatural alcohols ranging in length from five to eight carbons [144].

Engineering Tolerance to Inhibitory Compounds

As our repertoire of biologically-produced compounds increases, tolerance to high product titers becomes more important. Biofuels, such as ethanol and butanol, can inhibit biocatalyst growth, and therefore the tolerance of the biocatalyst needs to be improved [145–147]. As described above, our goal is to use lignocellulosic biomass as a substrate for production of commodity fuels and chemicals. Unfortunately, the processes used to convert biomass to soluble sugars also produce a mixture of minor products, such as furfural and acetic acid, that inhibit biocatalyst metabolism [148]. Although most of these inhibitors could be removed by detoxification [149], this additional process would increase operational cost. It is, thus, desirable to obtain microorganisms that are tolerant to these inhibitors and can directly ferment hemicellulose hydrolysate.

One approach to increasing tolerance is to understand the mechanism of inhibition. Transcriptome analysis has been used to probe the response to ethanol [145, 150], furfural [151], and butanol [147]. Another approach is to use directed evolution, as highlighted by the following example. Ethanologenic E. coli strain LY180 (a derivative of LY168 with restored lactose utilization and integration of an endoglucanase, and cellobiose utilization) was used as the chassis to select for furfural resistance through evolution [148]. The evolved strain, EMFR9, had significantly increased furfural resistance. Reverse engineering efforts, including transcriptome analysis, attributed furfural resistance to the silencing expression of several oxidoreductases. These oxidoreductases use NADPH for furfural reduction, depleting the available pools for biosynthesis. Thus furfural-mediated growth inhibition can be attributed to NADPH depletion [148], an insight that can be applied to other biocatalyst design projects.

PERSPECTIVES

Although many biocatalysts have been successfully redesigned for production of industrially important fuels and chemicals through traditional metabolic engineering, we are just beginning to see the potential of synthetic biology in this area. One of the foremost goals in our lab is the improvement of biocatalysts for biomass utilization. To attain this goal, tolerance to hydrolysate-derived inhibitors needs to be improved. For all applications, tolerance to high substrate and product titers is also important. This goal of redesigning a biocatalyst's phenotype, that is, tolerance, is not as clear as redesigning metabolism and a rational redesign strategy is particularly difficult when the mechanism of inhibition is not known.

As the understanding of our biocatalysts improves, particularly through reverse engineering of evolved strains, genome-scale models can be improved. Inclusion of kinetic and regulatory effects will also improve the accuracy and predictive power of these models. Note that some models have recently been developed that bypass the need for kinetic data, though [152]. Since enzymes are the major functional part performing the metabolic synthesis, improved protein engineering tools and new protein catalytic capability will aid in advancement of this field. It is important to generate high-quality protein mutagenesis libraries (relatively small libraries with a high diversity of enzymes) to facilitate efficient screening efforts [138]. Direct screening from metagenomic libraries of environmental samples can aid in isolation of enzymes with new functions, which cannot be obtained by the traditional strain isolation methods [153]. Enzymes can even be synthesized from scratch by a rational design strategy with computational aid [154]. Finally, new tools for better de novodesign of synthetic pathways need to be developed. Several databases, such as BNICE (Biochemical Network Integrated Computational Explorer) [155] and ReBiT (Retro-Biosynthesis Tool) [142], have already been established to facilitate identification of enzymes to construct a complete synthetic pathway for producing target compounds. It is important to establish guidelines, such as redox balance, energy production, and thermodynamic feasibility, to screen among these enormous pathways for the optimal routes.

By including synthetic biology tools in metabolic engineering projects, and vice versa, these two fields can significantly advance the replacement of petroleum-derived commodity products with those produced from bio renewable carbon and energy.

REFERENCES

1. L. R. Jarboe, T. B. Grabar, L. P. Yomano, K. T. Shanmugan, and L. O. Ingram, "Development of ethanologenic bacteria," Advances in Biochemical Engineering/Biotechnology, vol. 108, pp. 237–261, 2007.

2. D. G. Gibson, G. A. Benders, C. Andrews-Pfannkoch, et al., "Complete chemical synthesis, assembly, and cloning of a Mycoplasma genitalium genome," Science, vol. 319, no. 5867, pp. 1215–1220, 2008.

3. C. Laitigue, S. Vashee, M. A. Algire, et al., "Creating bacterial strains from genomes that have been cloned and engineered in yeast," Science, vol. 325, no. 5948, pp. 1693–1696, 2009.

4. J. E. Bailey, "Toward a science of metabolic engineering," Science, vol. 252, no. 5013, pp. 1668–1675, 1991.

5. G. Stephanopoulos and J. J. Vallino, "Network rigidity and metabolic engineering in metabolite overproduction," Science, vol. 252, no. 5013, pp. 1675–1681, 1991.

6. G. N. Stephanopoulos, A. A. Aristidou, and J. Nielsen, Metabolic Engineering: Principles and Methodologies, Academic Press, San Diego, Calif, USA, 1998.

7. P. Ball, "Synthetic biology: starting from scratch," Nature, vol. 431, no. 7009, pp. 624–626, 2004.

8. S. A. Benner and A. M. Sismour, "Synthetic biology," Nature Reviews Genetics, vol. 6, no. 7, pp. 533–543, 2005.

9.   R. McDaniel and R. Weiss, "Advances in synthetic biology: on the path from prototypes to applications," Current Opinion in Biotechnology, vol. 16, no. 4, pp. 476–483, 2005.

10. J. Pleiss, "The promise of synthetic biology," Applied Microbiology and Biotechnology, vol. 73, no. 4, pp. 735–739, 2006.

11. L. Serrano, "Synthetic biology: promises and challenges," Molecular Systems Biology, vol. 3, article 158, 2007.

12. J. D. Keasling, "Synthetic biology for synthetic chemistry," ACS Chemical Biology, vol. 3, no. 1, pp. 64–76, 2008.

13. A. C. Forster and G. M. Church, "Synthetic biology projects in vitro," Genome Research, vol. 17, no. 1, pp. 1–6, 2007.

14. D. Greber and M. Fussenegger, "Mammalian synthetic biology: engineering of sophisticated gene networks," Journal of Biotechnology, vol. 130, no. 4, pp. 329–345, 2007.

15. W. Weber and M. Fussenegger, "The impact of synthetic biology on drug discovery," Drug Discovery Today, vol. 14, no. 19-20, pp. 956–963, 2009.

16. C. E. French, "Synthetic biology and biomass conversion: a match made in heaven?" Journal of the Royal Society Interface, vol. 6, supplement 4, pp. S547–S558, 2009.

17. S. K. Lee, H. Chou, T. S. Ham, T. S. Lee, and J. D. Keasling, "Metabolic engineering of microorganisms for biofuels production: from bugs to synthetic biology to fuels," Current Opinion in Biotechnology, vol. 19, no. 6, pp. 556–563, 2008.

18. S. Picataggio, "Potential impact of synthetic biology on the development of microbial systems for the production of renewable fuels and chemicals," Current Opinion in Biotechnology, vol. 20, no. 3, pp. 325–329, 2009.

19. A. M. Feist, C. S. Henry, J. L. Reed, et al., "A genomescale metabolic reconstruction for Escherichia coli K-12 MG1655 that accounts for 1260 ORFs and thermodynamic information," Molecular Systems Biology, vol. 3, article 121, 2007.

20. I. M. Keseler, C. Bonavides-Mart'ınez, J. Collado-Vides, et al., "EcoCyc: a comprehensive view of Escherichia coli biology," Nucleic Acids Research, vol. 37, supplement 1, pp. D464–D470, 2009.

21. D. Endy, "Foundations for engineering biology," Nature, vol. 438, no. 7067, pp. 449–453, 2005.

22. T. B. Grabar, S. Zhou, K. T. Shanmugam, L. P. Yomano, and L. O. Ingram, "Methylglyoxal bypass identified as source of chiral contamination in L(+) and D(-)-lactate fermentations by recombinant Escherichia coli," Biotechnology Letters, vol. 28, no. 19, pp. 1527–1535, 2006.

23. T. B. Causey, S. Zhou, K. T. Shanmugam, and L. O. Ingram, "Engineering the metabolism of Escherichia coli W3110 for the conversion of sugar to redox-neutral and oxidized products: homoacetate production," Proceedings of the National Academy of Sciences of the United States of America, vol. 100, no. 3, pp. 825–832, 2003.

24. K. Jantama, X. Zhang, J. C. Moore, K. T. Shanmugam, S. A. Svoronos, and L. O. Ingram, "Eliminating side products and increasing succinate yields in engineered strains of Escherichia coli C," Biotechnology and Bioengineering, vol. 101, no. 5, pp. 881–893, 2008.

25. L. P. Yomano, S. W. York, K. T. Shanmugam, and L. O. Ingram, "Deletion of methylglyoxal synthase gene (mgsA) increased sugar co-metabolism in ethanol-producing Escherichia coli," Biotechnology Letters, vol. 31, no. 9, pp. 1389–1398, 2009.

26. P. C. Cirino, J. W. Chin, and L. O. Ingram, "Engineering Escherichia coli for xylitol production from glucose-xylose mixtures," Biotechnology and Bioengineering, vol. 95, no. 6, pp. 1167–1176, 2006.

27. X. Zhang, K. Jantama, J. C. Moore, K. T. Shanmugam, and L. O. Ingram, "Production of L-alanine by metabolically engineered Escherichia coli," Applied Microbiology and Biotechnology, vol. 77, no. 2, pp. 355–366, 2007.

28. R. S. Senger and E. T. Papoutsakis, "Genome-scale model for Clostridium acetobutylicum—part I: metabolic network resolution and analysis," Biotechnology and Bioengineering, vol. 101, no. 5, pp. 1036–1052, 2008.

29. K. R. Kjeldsen and J. Nielsen, "In silico genome-scale reconstruction and validation of the corynebacterium glutamicum metabolic network," Biotechnology and Bioengineering, vol. 102, no. 2, pp. 583–597, 2009.

30. J. Forster, I. Famili, P. Fu, et al., "Genome-scale reconstruction of the Saccharomyces cerevisiae metabolic network," Genome Research, vol. 13, no. 2, pp. 244–253, 2003.

31. M. R. Andersen, M. L. Nielsen, and J. Nielsen, "Metabolic model integration of the bibliome, genome, metabolome and reactome of Aspergillus niger," Molecular Systems Biology, vol. 4, article 178, 2008.

32. F. R. Blattner, et al., "The complete genome sequence of Escherichia coli K-12," Science, vol. 277, no. 5331, pp. 1453–1474, 1997.

33. R. U. Ibarra, J. S. Edwards, and B. O. Palsson, "Escherichia coli K-12 undergoes adaptive evolution to achieve in silico predicted optimal growth," Nature, vol. 420, no. 6912, pp. 186–189, 2002.

34. A. P. Bauer, S. M. Dieckmann, W. Ludwig, and K.-H. Schleifer, "Rapid identification of Escherichia coli safety and laboratory strain lineages based on Multiplex-PCR," FEMS Microbiology Letters, vol. 269, no. 1, pp. 36–40, 2007.

35. S. Zhou, L. P. Yomano, K. T. Shanmugam, and L. O. Ingram, "Fermentation of 10% (w/v) sugar to D: (−)-lactate by engineered Escherichia coli B," Biotechnology Letters, vol. 27, no. 23-24, pp. 1891–1896, 2005.

36. M. Moniruzzaman, X. Lai, S. W. York, and L. O. Ingram, "Isolation and molecular characterization of high-performance cellobiose- fermenting spontaneous mutants of ethanologenic Escherichia coli KO11 containing the Klebsiella oxytoca

casAB operon," Applied and Environmental Microbiology, vol. 63, no. 12, pp. 4633–4637, 1997.

37. G. Posfai, G. Plunkett III, T. Feher, et al., "Emergent ´properties of reduced-genome Escherichia coli," Science, vol. 312, no. 5776, pp. 1044–1046, 2006.

38. M. Durot, P.-Y. Bourguignon, and V. Schachter, "Genomescale models of bacterial metabolism: reconstruction and applications," FEMS Microbiology Reviews, vol. 33, no. 1, pp. 164–190, 2009.

39. N. D. Price, J. A. Papin, C. H. Schilling, and B. O. Palsson, "Genome-scale microbial in silico models: the constraintsbased approach," Trends in Biotechnology, vol. 21, no. 4, pp. 162–169, 2003.

40. P. Pharkya, A. P. Burgard, and C. D. Maranas, "OptStrain: a computational framework for redesign of microbial production systems," Genome Research, vol. 14, no. 11, pp. 2367– 2376, 2004.

41. A. P. Burgard, P. Pharkya, and C. D. Maranas, "OptKnock: a bilevel programming framework for identifying gene knockout strategies for microbial strain optimization," Biotechnology and Bioengineering, vol. 84, no. 6, pp. 647–657, 2003.

42. H. Alper, Y.-S. Jin, J. F. Moxley, and G. Stephanopoulos, "Identifying gene targets for the metabolic engineering of lycopene biosynthesis in Escherichia coli," Metabolic Engineering, vol. 7, no. 3, pp. 155–164, 2005.

43. C. Bro and J. Nielsen, "Impact of 'ome' analyses on inverse metabolic engineering," Metabolic Engineering, vol. 6, no. 3, pp. 204–211, 2004.

44. S. Y. Lee, D. Y. Lee, and T. Y. Kim, "Systems biotechnology for strain improvement," Trends in Biotechnology, vol. 23, no. 7, pp. 349–358, 2005.

45. T. Hermann, "Using functional genomics to improve productivity in the manufacture of industrial biochemicals," Current Opinion in Biotechnology, vol. 15, no. 5, pp. 444–448, 2004.

46. J. H. Choi, S. J. Lee, and S. Y. Lee, "Enhanced production of insulin-like growth factor I fusion protein in Escherichia coli by coexpression of the down-regulated genes identified by transcriptome profiling," Applied and Environmental Microbiology, vol. 69, no. 8, pp. 4737–4742, 2003.

47. M.-J. Han, K. J. Jeong, J.-S. Yoo, and S. Y. Lee, "Engineering Escherichia coli for increased productivity of serinerich proteins based on proteome profiling," Applied and Environmental Microbiology, vol. 69, no. 10, pp. 5772–5781, 2003.

48. B. F. Pfleger, D. J. Pitera, C. D. Smolke, and J. D. Keasling, "Combinatorial engineering of intergenic regions in operons tunes expression of multiple genes," Nature Biotechnology, vol. 24, no. 8, pp. 1027–1032, 2006.

49. S. A. Underwood, M. L. Buszko, K. T. Shanmugam, and L. O. Ingram, "Flux through citrate synthase limits the growth of ethanologenic Escherichia coli KO11 during xylose fermentation," Applied and Environmental Microbiology, vol. 68, no. 3, pp. 1071–1081, 2002.

50. M. Askenazi, E. M. Driggers, D. A. Holtzman, et al., "Integrating transcriptional and metabolite profiles to direct the engineering of lovastatin-producing fungal strains," Nature Biotechnology, vol. 21, no. 2, pp. 150–156, 2003.

51. J. Ohnishi, S. Mitsuhashi, M. Hayashi, et al., "A novel methodology employing Corynebacterium glutamicum genome information to generate a new L-lysine-producing mutant," Applied Microbiology and Biotechnology, vol. 58, no. 2, pp. 217–223, 2002.

52. F. Martinez-Morales, A. C. Borges, A. Martinez, K. T. Shanmugam, and L. O. Ingram, "Chromosomal integration of heterologous DNA in Escherichia coli with precise removal of markers and replicons used during construction," Journal of Bacteriology, vol. 181, no. 22, pp. 7143–7148, 1999.

53. K. A. Datsenko and B. L. Wanner, "One-step inactivation of chromosomal genes in Escherichia coli K-12 using PCR

products," Proceedings of the National Academy of Sciences of the United States of America, vol. 97, no. 12, pp. 6640–6645, 2000.

54. G. Posfai, M. D. Koob, H. A. Kirkpatrick, and F. R. Blattner, "Versatile insertion plasmids for targeted genome manipulations in bacteria: isolation, deletion, and rescue of the pathogenicity island LEE of the Escherichia coli O157:H7 genome," Journal of Bacteriology, vol. 179, no. 13, pp. 4426–4428, 1997.

55. P. Gay, D. Le Coq, M. Steinmetz, et al., "Positive selection procedure for entrapment of insertion sequence elements in gram-negative bacteria," Journal of Bacteriology, vol. 164, no. 2, pp. 918–921, 1985.

56. G. de la Cueva-Mendez and B. Pimentel, "Gene and cell survival: lessons from prokaryotic plasmid R1," The EMBO Reports, vol. 8, no. 5, pp. 458–464, 2007.

57. S. Zhou, F. C. Davis, and L. O. Ingram, "Gene integration and expression and extracellular secretion of Erwinia chrysanthemi endoglucanase CelY (celY) and CelZ (celZ) in ethanologenic Klebsiella oxytoca P2," Applied and Environmental Microbiology, vol. 67, no. 1, pp. 6–14, 2001.

58. P. R. Jensen and K. Hammer, "The sequence of spacers between the consensus sequences modulates the strength of prokaryotic promoters," Applied and Environmental Microbiology, vol. 64, no. 1, pp. 82–87, 1998.

59. H. Alper, C. Fischer, E. Nevoigt, and G. Stephanopoulos, "Tuning genetic control through promoter engineering," Proceedings of the National Academy of Sciences of the United States of America, vol. 102, no. 36, pp. 12678–12683, 2005.

60. I. Meynial-Salles, M. A. Cervin, and P. Soucaille, "New tool for metabolic pathway engineering in Escherichia coli: onestep method to modulate expression of chromosomal genes," Applied and Environmental Microbiology, vol. 71, no. 4, pp. 2140–2144, 2005.

61. C. Solem and P. R. Jensen, "Modulation of gene expression made easy," Applied and Environmental Microbiology, vol. 68, no. 5, pp. 2397–2403, 2002.

62. T. A. Carrier and J. D. Keasling, "Library of synthetic 5' secondary structures to manipulate mRNA stability in Escherichia coli," Biotechnology Progress, vol. 15, no. 1, pp. 58–64, 1999.

63. Y. S. Park, S. W. Seo, S. Hwang, et al., "Design of 5'-untranslated region variants for tunable expression in Escherichia coli," Biochemical and Biophysical Research Communications, vol. 356, no. 1, pp. 136–141, 2007.

64. F. J. Isaacs, D. J. Dwyer, C. Ding, D. D. Pervouchine, C. R. Cantor, and J. J. Collins, "Engineered riboregulators enable post-transcriptional control of gene expression," Nature Biotechnology, vol. 22, no. 7, pp. 841–847, 2004.

65. S. Jana and J. K. Deb, "Strategies for efficient production of heterologous proteins in Escherichia coli," Applied Microbiology and Biotechnology, vol. 67, no. 3, pp. 289–298, 2005.

66. H. Alper, J. Moxley, E. Nevoigt, G. R. Fink, and G. Stephanopoulos, "Engineering yeast transcription machinery for improved ethanol tolerance and production," Science, vol. 314, no. 5805, pp. 1565–1568, 2006.

67. H. Alper and G. Stephanopoulos, "Global transcription machinery engineering: a new approach for improving cellular phenotype," Metabolic Engineering, vol. 9, no. 3, pp. 258–267, 2007.

68. I. P. Petrounia and F. H. Arnold, "Designed evolution of enzymatic properties," Current Opinion in Biotechnology, vol. 11, no. 4, pp. 325–330, 2000.

69. H. Zhao, L. Giver, Z. Shao, J. A. Affholter, and F. H. Arnold, "Molecular evolution by staggered extension process (StEP) in vitro recombination," Nature Biotechnology, vol. 16, no. 3, pp. 258–261, 1998.

70.   A. Crameri, S. A. Raillard, E. Bermudez, and W. P. C. Stemmer, "DNA shuffling of a family of genes from diverse species accelerates directed evolution," Nature, vol. 391, no. 6664, pp. 288–291, 1998.

71.   C. Wang, M.-K. Oh, and J. C. Liao, "Directed evolution of metabolically engineered Escherichia coli for carotenoid production," Biotechnology Progress, vol. 16, no. 6, pp. 922–926, 2000.

72.   P. C. Cirino and F. H. Arnold, "Protein engineering of oxygenases for biocatalysis," Current Opinion in Chemical Biology, vol. 6, no. 2, pp. 130–135, 2002.

73.   H. H. Liao, "Alanine 2,3-aminomutase," US patent 7309597, 2007.

74.   M. M. Altamirano, J. M. Blackburn, C. Aguayo, and A. R. Fersht, "Directed evolution of new catalytic activity using the α/β-barrel scaffold," Nature, vol. 403, no. 6770, pp. 617–622, 2000.

75.   M. Leisola and O. Turunen, "Protein engineering: opportunities and challenges," Applied Microbiology and Biotechnology, vol. 75, no. 6, pp. 1225–1232, 2007.

76.   A. Korkegian, M. E. Black, D. Baker, and B. L. Stoddard, "Computational thermostabilization of an enzyme," Science, vol. 308, no. 5723, pp. 857–860, 2005.

77.   R. Machielsen, L. L. Looger, J. Raedts, et al., "Cofactor engineering of Lactobacillus brevis alcohol dehydrogenase by computational design," Engineering in Life Sciences, vol. 9, no. 1, pp. 38–44, 2009.

78.   J. D. Bloom, M. M. Meyer, P. Meinhold, C. R. Otey, D. MacMillan, and F. H. Arnold, "Evolving strategies for enzyme engineering," Current Opinion in Structural Biology, vol. 15, no. 4, pp. 447–452, 2005.

79.   R. A. Chica, N. Doucet, and J. N. Pelletier, "Semi-rational approaches to engineering enzyme activity: combining the benefits of directed evolution and rational design," Current Opinion in Biotechnology, vol. 16, no. 4, pp. 378–384, 2005.

80. O. Spadiut, I. Pisanelli, T. Maischberger, et al., "Engineering of pyranose 2-oxidase: improvement for biofuel cell and food applications through semi-rational protein design," Journal of Biotechnology, vol. 139, no. 3, pp. 250–257, 2009.

81. Q. Wang, A. R. Parrish, and L. Wang, "Expanding the genetic code for biological studies," Chemistry and Biology, vol. 16, no. 3, pp. 323–336, 2009.

82. Y. Lu, "Design and engineering of metalloproteins containing unnatural amino acids or non-native metal-containing cofactors," Current Opinion in Chemical Biology, vol. 9, no. 2, pp. 118–126, 2005.

83. J. C. Jackson, S. P. Duffy, K. R. Hess, and R. A. Mehl, "Improving nature's enzyme active site with genetically encoded unnatural amino acids," Journal of the American Chemical Society, vol. 128, no. 34, pp. 11124–11127, 2006.

84. D. B. Walker, G. Joshi, and A. P. Davis, "Progress in biomimetic carbohydrate recognition," Cellular and Molecular Life Sciences, vol. 66, no. 19, pp. 3177–3191, 2009.

85. M. P. Cashion and T. E. Long, "Biomimetic design and performance of polymerizable lipids," Accounts of Chemical Research, vol. 42, no. 8, pp. 1016–1025, 2009.

86. S. S. Fong, A. P. Burgard, C. D. Herring, et al., "In silico design and adaptive evolution of Escherichia coli for production of lactic acid," Biotechnology and Bioengineering, vol. 91, no. 5, pp. 643–648, 2005.

87. X. Zhang, K. Jantama, J. C. Moore, L. R. Jarboe, K. T. Shanmugam, and L. O. Ingram, "Metabolic evolution of energy-conserving pathways for succinate production in Escherichia coli," Proceedings of the National Academy of Sciences of the United States of America, vol. 106, no. 48, pp. 20180–20185, 2010.

88. J. G. Zeikus, M. K. Jain, and P. Elankovan, "Biotechnology of succinic acid production and markets for derived industrial products," Applied Microbiology and Biotechnology, vol. 51, no. 5, pp. 545–552, 1999.

89. J. B. McKinlay, C. Vieille, and J. G. Zeikus, "Prospects for a bio-based succinate industry," Applied Microbiology and Biotechnology, vol. 76, no. 4, pp. 727–740, 2007.

90. N. S. Samuelov, R. Lamed, S. Lowe, and I. G. Zeikus, "Influence of CO2-HCO3- levels and ph on growth, succinate production, and enzyme activities of Anaerobiospirillum succiniciproducens," Applied and Environmental Microbiology, vol. 57, no. 10, pp. 3013–3019, 1991.

91. M. J. Vander Werf, M. V. Guettler, M. K. Jain, and J. G. Zeikus, "Environmental and physiological factors affecting the succinate product ratio during carbohydrate fermentation by Actinobacillus sp. 130Z," Archives of Microbiology, vol. 167, no. 6, pp. 332–342, 1997.

92. P. Lee, S. Lee, S. Hong, and H. Chang, "Isolation and characterization of a new succinic acid-producing bacterium, Mannheimia succiniciproducens MBEL55E, from bovine rumen," Applied Microbiology and Biotechnology, vol. 58, no. 5, pp. 663–668, 2002.

93. D. G. Fraenkel, "Glycolysis," in Escherichia coli and Salmonella: Cellular and Molecular Biology, F. C. Neidhardt, Ed., ASM Press, Washington, DC, USA, 1996.

94. K. Jantama, M. J. Haupt, S. A. Svoronos, et al., "Combining metabolic engineering and metabolic evolution to develop nonrecombinant strains of Escherichia coli C that produce succinate and malate," Biotechnology and Bioengineering, vol. 99, no. 5, pp. 1140–1153, 2008.

95. M.-K. Oh, L. Rohlin, K. C. Kao, and J. C. Liao, "Global expression profiling of acetate-grown Escherichia coli," Journal of Biological Chemistry, vol. 277, no. 15, pp. 13175–13183, 2002.

96. K. C. Kao, L. M. Tran, and J. C. Liao, "A global regulatory role of gluconeogenic genes in Escherichia coli revealed by transcriptome network analysis," Journal of Biological Chemistry, vol. 280, no. 43, pp. 36079–36087, 2005.

97. R. Patnaik, W. D. Roof, R. F. Young, and J. C. Liao, "Stimulation of glucose catabolism in Escherichia coli by a potential futile cycle," Journal of Bacteriology, vol. 174, no. 23, pp. 7527–7532, 1992.

98. X. Zhang, K. Jantama, K. T. Shanmugam, and L. O. Ingram, "Reengineering Escherichia coli for succinate production in mineral salts medium," Applied and Environmental Microbiology, vol. 75, no. 24, pp. 7807–7813, 2009.

99. S. S. Ray and M. Bousmina, "Biodegradable polymers and their layered silicate nano composites: in greening the 21st century materials world," Progress in Materials Science, vol. 50, no. 8, pp. 962–1079, 2005.

100. A. K. Agrawal and R. Bhalla, "Advances in the production of poly(lactic acid) fibers. A review," Journal of Macromolecular Science-Polymer Reviews C, vol. 43, no. 4, pp. 479–503, 2003.

101. S. Benthin and J. Villadsen, "Production of optically pure Dlactate by lactobacillus bulgaricus and purification by crystallisation and liquid/liquid extraction," Applied Microbiology and Biotechnology, vol. 42, no. 6, pp. 826–829, 1995.

102. A. Demirci and A. L. Pometto, "Enhanced production of D(-)-lactic acid by mutants of lactobacillus-delbrueckii ATCC-9649," Journal of Industrial Microbiology, vol. 11, no. 1, pp. 23–28, 1992.

103. D. G. Fraenkel, "Glycolysis," in Escherichia coli and Salmonella: Cellular and Molecular Biology, F. C. Neidhardt, Ed., vol. 1, chapter 14, ASM Press, Washington DC, USA, 2nd edition, 1996.

104. S. Zhou, T. B. Causey, A. Hasona, K. T. Shanmugam, and L. O. Ingram, "Production of optically pure D-lactic acid in mineral salts medium by metabolically engineered Escherichia coli W3110," Applied and Environmental Microbiology, vol. 69, no. 1, pp. 399–407, 2003.

105. S. Zhou, K. T. Shanmugam, L. P. Yomano, T. B. Grabar, and L. O. Ingram, "Fermentation of 12% (w/v) glucose to 1.2 M lactate by Escherichia coli strain SZ194 using mineral salts medium," Biotechnology Letters, vol. 28, no. 9, pp. 663–670, 2006.

106. M. Cheryan, S. Parekh, M. Shah, and K. Witjitra, "Production of acetic acid by Clostridium thermoaceticum," Advances in Applied Microbiology, vol. 43, pp. 1–33, 1997.

107. C. Berraud, "Production of highly concentrated vinegar in fed-batch culture," Biotechnology Letters, vol. 22, no. 6, pp. 451–454, 2000.

108. S. N. Freer, "Acetic acid production by Dekkera/Brettanomyces yeasts," World Journal of Microbiology and Biotechnology, vol. 18, no. 3, pp. 271–275, 2002.

109. S. Y. Lee, J. H. Park, S. H. Jang, L. K. Nielsen, J. Kim, and K. S. Jung, "Fermentative butanol production by clostridia," Biotechnology and Bioengineering, vol. 101, no. 2, pp. 209–228, 2008.

110. L. D. Mermelstein, E. T. Papoutsakis, D. J. Petersen, and G. N. Bennett, "Metabolic engineering of Clostridium acetobutylicum ATCC 824 for increased solvent production by enhancement of acetone formation enzyme activities using a synthetic acetone operon," Biotechnology and Bioengineering, vol. 42, no. 9, pp. 1053–1060, 1993.

111. L. Harris, L. Blank, R. P. Desai, N. E. Welker, and E. T. Papoutsakis, "Fermentation characterization and flux analysis of recombinant strains of Clostridium acetobutylicum with an inactivated solR gene," Journal of Industrial Microbiology and Biotechnology, vol. 27, no. 5, pp. 322–328, 2001.

112. R. V. Nair, E. M. Green, D. E. Watson, G. N. Bennett, and E. T. Papoutsakis, "Regulation of the sol locus genes for butanol and acetone formation in Clostridium acetobutylicum ATCC 824 by a putative transcriptional repressor," Journal of Bacteriology, vol. 181, no. 1, pp. 319–330, 1999.

113. R. P. Desai and E. T. Papoutsakis, "Antisense RNA strategies for metabolic engineering of Clostridium acetobutylicum," Applied and Environmental Microbiology, vol. 65, no. 3, pp. 936–945, 1999.

114. N. E. Altaras and D. C. Cameron, "Enhanced production of (R)-1,2-propanediol by metabolically engineered Escherichia coli," Biotechnology Progress, vol. 16, no. 6, pp. 940–946, 2000.

115. P. Soucaille, I. Meynial-Salles, F. Voelker, and R. Figge, "Microorganisms and methods for prodcution of 1,2-propanediol and acetol," WO, 2008, 2008/116853.

116. J. H. Park, K. H. Lee, T. Y. Kim, and S. Y. Lee, "Metabolic engineering of Escherichia coli for the production of L-valine based on transcriptome analysis and in silico gene knockout simulation," Proceedings of the National Academy of Sciences of the United States of America, vol. 104, no. 19, pp. 7797–7802, 2007.

117. K. H. Lee, J. H. Park, T. Y. Kim, H. U. Kim, and S. Y. Lee, "Systems metabolic engineering of Escherichia coli for L-threonine production," Molecular Systems Biology, vol. 3, article 149, 2007.

118. L. O. Ingram, P. F. Gomez, X. Lai, et al., "Metabolic engineering of bacteria for ethanol production," Biotechnology and Bioengineering, vol. 58, no. 2-3, pp. 204–214, 1998.

119. Y. Kim, L. O. Ingram, and K. T. Shanmugam, "Construction of an Escherichia coli K-12 mutant for homoethanologenic fermentation of glucose or xylose without foreign genes," Applied and Environmental Microbiology, vol. 73, no. 6, pp. 1766–1771, 2007.

120. L. O. Ingram, T. Conway, D. P. Clark, G. W. Sewell, and J. F. Preston, "Genetic engineering of ethanol production in Escherichia coli," Applied and Environmental Microbiology, vol. 53, no. 10, pp. 2420–2425, 1987.

121. K. Ohta, D. S. Beall, J. P. Mejia, K. T. Shanmugam, and L. O. Ingram, "Genetic improvement of Escherichia coli for

ethanol production: chromosomal integration of Zymomonas mobilis genes encoding pyruvate decarboxylase and alcohol dehydrogenase II," Applied and Environmental Microbiology, vol. 57, no. 4, pp. 893–900, 1991.

122. L. P. Yomano, S. W. York, S. Zhou, K. T. Shanmugam, and L. O. Ingram, "Re-engineering Escherichia coli for ethanol production," Biotechnology Letters, vol. 30, no. 12, pp. 2097– 2103, 2008.

123. A. Hasona, S. W. York, L. P. Yomano, L. O. Ingram, and K. T. Shanmugam, "Decreasing the level of ethyl acetate in ethanolic fermentation broths of Escherichia coli KO11 by expression of Pseudomonas putida estZ esterase," Applied and Environmental Microbiology, vol. 68, no. 6, pp. 2651– 2659, 2002.

124. K. Hofvendahl and B. Hahn-Hagerdal, "Factors affecting the fermentative lactic acid production from renewable resources," Enzyme and Microbial Technology, vol. 26, no. 2– 4, pp. 87–107, 2000.

125. S. D. Zhou, K. T. Shanmugam, and L. O. Ingram, "Functional replacement of the Escherichia coli D-(-)-lactate dehydrogenase gene (ldhA) with the L-(+)-lactate dehydrogenase gene (ldhL) from Pediococcus acidilactici," Applied and Environmental Microbiology, vol. 69, no. 4, pp. 2237–2244, 2003.

126. J. C. Parajo, H. Dom ´ ´inguez, and J. M. Dom´inguez, "Biotechnological production of xylitol—part 1: interest of xylitol and fundamentals of its biosynthesis," Bioresource Technology, vol. 65, no. 3, pp. 191–201, 1998.

127. T. Werpy and G. Petersen, Eds., Top Value Added Chemicals from Biomass: Volume I—Results of Screening for Potential Candidates from Sugars and Synthesis Gas, Pacific Northwest National Laboratory and National Renewable Energy Laboratory, 2004.

128. T. B. Kim and D. K. Oh, "Xylitol production by Candida tropicalis in a chemically defined medium," Biotechnology Letters, vol. 25, no. 24, pp. 2085–2088, 2003.

129. J. W. Chin, R. Khankal, C. A. Monroe, C. D. Maranas, and P. C. Cirino, "Analysis of NADPH supply during xylitol production by engineered Escherichia coli," Biotechnology and Bioengineering, vol. 102, no. 1, pp. 209–220, 2009.

130. P. Hols, M. Kleerebezem, A. N. Schanck, et al., "Conversion of Lactococcus lactis from homolactic to homoalanine fermentation through metabolic engineering," Nature Biotechnology, vol. 17, no. 6, pp. 588–592, 1999.

131. M. Ikeda, "Amino acid production processes," Advances in Biochemical Engineering/Biotechnology, vol. 79, pp. 1–35, 2003.

132. C. E. Nakamura and G. M. Whited, "Metabolic engineering for the microbial production of 1,3-propanediol," Current Opinion in Biotechnology, vol. 14, no. 5, pp. 454–459, 2003.

133. T. Hanai, S. Atsumi, and J. C. Liao, "Engineered synthetic pathway for isopropanol production in Escherichia coli," Applied and Environmental Microbiology, vol. 73, no. 24, pp. 7814–7818, 2007.

134. D.-K. Ro, E. M. Paradise, M. Quellet, et al., "Production of the antimalarial drug precursor artemisinic acid in engineered yeast," Nature, vol. 440, no. 7086, pp. 940–943, 2006.

135. Y. Yan, A. Kohli, and M. A. G. Koffas, "Biosynthesis of natural flavanones in Saccharomyces cerevisiae," Applied and Environmental Microbiology, vol. 71, no. 9, pp. 5610–5613, 2005.

136. E. Leonard, Y. Yan, and M. A. G. Koffas, "Functional expression of a P450 flavonoid hydroxylase for the biosynthesis of plant-specific hydroxylated flavonols in Escherichia coli," Metabolic Engineering, vol. 8, no. 2, pp. 172–181, 2006.

137. E. Leonard, K. H. Lim, P.-N. Saw, and M. A. G. Koffas, "Engineering central metabolic pathways for high-level flavonoid production in Escherichia coli," Applied and Environmental Microbiology, vol. 73, no. 12, pp. 3877–3886, 2007.

138. A. L. de Boer and C. Schmidt-Dannert, "Recent efforts in engineering microbial cells to produce new chemical compounds," Current Opinion in Chemical Biology, vol. 7, no. 2, pp. 273–278, 2003.

139. C. Schmidt-Dannert, D. Umeno, and F. H. Arnold, "Molecular breeding of carotenoid biosynthetic pathways," Nature Biotechnology, vol. 18, no. 7, pp. 750–753, 2000.

140. G. Sandmann, "Combinatorial biosynthesis of carotenoids in a heterologous host: a powerful approach for the biosynthesis of novel structures," ChemBioChem, vol. 3, no. 7, pp. 629–635, 2002.

141. M. Albrecht, S. Takaichi, S. Steiger, Z.-Y. Wang, and G. Sandmann, "Novel hydroxycarotenoids with improved antioxidative properties produced by gene combination in Escherichia coli," Nature Biotechnology, vol. 18, no. 8, pp. 843– 846, 2000.

142. K. L. J. Prather and C. H. Martin, "De novo biosynthetic pathways: rational design of microbial chemical factories," Current Opinion in Biotechnology, vol. 19, no. 5, pp. 468–474, 2008.

143. S. Atsumi, T. Hanai, and J. C. Liao, "Non-fermentative pathways for synthesis of branched-chain higher alcohols as biofuels," Nature, vol. 451, no. 7174, pp. 86–89, 2008.

144. K. Zhang, M. R. Sawaya, D. S. Eisenberg, and J. C. Liao, "Expanding metabolism for biosynthesis of nonnatural alcohols," Proceedings of the National Academy of Sciences of the United States of America, vol. 105, no. 52, pp. 20653–20658, 2008.

145. R. Gonzalez, H. Tao, J. E. Purvis, S. W. York, K. T. Shanmugam, and L. O. Ingram, "Gene array-based identi- fication of changes that contribute to ethanol tolerance In ethanologenic Escherichia coli: comparison of KO11 (parent) to LY01 (resistant mutant)," Biotechnology Progress, vol. 19, no. 2, pp. 612–623, 2003.

146. L. P. Yomano, S. W. York, and L. O. Ingram, "Isolation and characterization of ethanol-tolerant mutants of Escherichia coli KO11 for fuel ethanol production," Journal of Industrial Microbiology and Biotechnology, vol. 20, no. 2, pp. 132–138, 1998.

147. M. P. Brynildsen and J. C. Liao, "An integrated network approach identifies the isobutanol response network of Escherichia coli," Molecular Systems Biology, vol. 5, article 277, 2009.

148. E. N. Miller, et al., "Silencing of NADPH-dependent oxidoreductase genes (yqhD and dkgA) in furfural-resistant ethanologenic Escherichia coli," Applied and Environmental Microbiology, vol. 75, no. 13, pp. 4315–4323, 2009.

149. A. Martinez, M. E. Rodriguez, M. L. Wells, S. W. York, J. F. Preston, and L. O. Ingram, "Detoxification of dilute acid hydrolysates of lignocellulose with lime," Biotechnology Progress, vol. 17, no. 2, pp. 287–293, 2001.

150. R. Gonzalez, H. Tao, K. T. Shanmugam, S.W. York, and L. O. Ingram, "Transcriptome analysis of ethanologenic Escherichia coli strains: tolerance to ethanol," in Proceedings of the 225th ACS National Meeting, p. U200, New Orleans, La, USA, March 2003.

151. E. N. Miller, L. R. Jarboe, P. C. Turner, et al., "Furfural inhibits growth by limiting sulfur assimilation in ethanologenic Escherichia coli strain LY180," Applied and Environmental Microbiology, vol. 75, no. 19, pp. 6132–6141, 2009.

152. L. M. Tran, M. L. Rizk, and J. C. Liao, "Ensemble modeling of metabolic networks," Biophysical Journal, vol. 95, no. 12, pp. 5606–5617, 2008.

153. P. D. Schloss and J. Handelsman, "Biotechnological prospects from metagenomics," Current Opinion in Biotechnology, vol. 14, no. 3, pp. 303–310, 2003.

154. L. Jiang, E. A. Althoff, F. R. Clemente, et al., "De novo computational design of retro-aldol enzymes," Science, vol. 319, no. 5868, pp. 1387–1391, 2008.

155. V. Hatzimanikatis, C. Li, J. A. Ionita, C. S. Henry, M. D. Jankowski, and L. Broadbelt, "Exploring the diversity of complex metabolic networks," Bioinformatics, vol. 21, no. 8, pp. 1603–1609, 2005.

Chapter **2**

A New Multichelating Acid System for High-Temperature Sandstone Reservoirs

Nianyin Li, Qian Zhang, Yongqing Wang,
Pingli Liu, and Liqiang Zhao

State Key Lab of Oil and Gas Reservoir Geology and Exploitation,
South West Petroleum University, Chengdu, Sichuan 610500, China

ABSTRACT

Sandstone reservoir acidizing is a complex and heterogeneous acid-rock reaction process. If improper acid treatment is implemented, further damage can be induced instead of removing the initial plug, particularly in high-temperature sandstone reservoirs. An efficient acid system is the key to successful acid treatment. High-temperature sandstone treatment with conventional mud acid system faces problems including high acid-rock reaction rate, short acid effective distance, susceptibility to secondary damage, and serious corrosion to pipelines. In this paper, a new multichelating acid system has been developed to overcome these shortcomings. The acid system is composed of ternary weak acid, organic phosphonic chelating agent, anionic polycarboxylic acid chelating dispersant, fluoride, and other assisted additives. Hydrogen ion slowly released by multistage ionization in ternary weak acid and organic phosphonic within the system decreases the concentration of HF to achieve retardation. Chelating agent and chelating dispersant within the system inhibited anodic and cathodic reaction, respectively, to protect the metal from corrosion, while chelating dispersant has great chelating ability on iron ions, restricting the depolarization reaction of ferric ion and metal. The synergic effect of chelating agent and chelating dispersant removes sulfate scale precipitation and inhibits or decreases potential precipitation such as CaF_2, silica gel, and fluosilicate. Mechanisms of retardation, corrosion-inhibition, and scale-removing features have been discussed and evaluated with laboratory tests. Test results indicate that this novel acid system has good overall performance, addressing the technical problems and improving the acidizing effect as well for high-temperature sandstone.

INTRODUCTION

Sandstone matrix acidizing is one of the essential technical strategies to maintain or increase the productivity of hydrocarbon wells or the injectivity of water injectors. With the flow and reaction of acid in intergranular porosity and cavity, the near wellbore region damaged by drilling, completion, workover, or injection of water can be removed to recover or increase the production. The acid-rock reaction in sandstone matrix is a fairly complex process, including the chemical reaction between a variety of minerals and hydrofluoric acid, which happens most in porous media, and is considered as multiphase "heterogeneity" reaction. Since there are different minerals reaction rate with acids and difficulty in measurement, the reaction process is beyond accurate prediction. Therefore, improper acidizing treatment not only cannot remove the original plugs, but also bring further damage to reservoir [1–5].

Stimulation effect is highly dependent on acid systems. To implement a successful acid treatment, an optimized acid system considering the formation characteristics and the function and performance of both acids and additives should be applied to meet requirements of the treatment.

In high temperature conditions, regular mud acid reacts rapidly, with limited effective distance, and untouched to the deep reservoir. Moreover, regular mud acid could induce massive secondary precipitation and heavy corrosion on downhole tubulars and devices. Corrosion inhibitors have been used with acids for decades but their function is limited in some extreme conditions, especially under high temperature circumstances. Corrosion inhibitors can be classified into 5 categories: amino polycarboxylic type, hydroxy carboxylic type, amino type, organic phosphate type, and polycarboxylic type. The first use of chelating agent was to dissolve carbonate in 1973 [6]. And in 1985, Tyler et al. first used EDTA in sandstone reservoir [7]. Several of researches have proved the great advantage of chelating agent in acid treatment of sandstone reservoir, leading to wide use like EDTA, HEDTA, NTA, and so forth [8]. Al-Harbi et al. introduced the application of chelating agents to sandstone acid systems in 2013 [9]. For acidizing treatment in high-temperature sandstone reservoir, chelating agents use is the future for improving treatment effectively [10–17]. Due to the limitations of pros and cons of single chelating agent usage, this paper introduces a synergic affected optimized mixture of chelating agent and dispersant, a high-temperature multichelating acid system with satisfactory effect on retardation, corrosion inhibition, scale inhibition, and scale removing through careful designs and detailed characteristic evaluation.

DESIGN OF HIGH TEMPERATURE MULTICHELATING ACID SYSTEM

The multichelating acid system consists of H_3PO_4, SAV-1, SAV-2, NH_4F, and additives, with the organic phosphoric chelating agent SAV-1 and the anionic poly carboxylic acid chelating dispersant SAV-2. Using the weak acid to slow the releasing rate of hydrogen ion can decrease the HF concentration for retardation. Chelating agents and dispersants are utilized to prevent or decrease the generation of secondary precipitation. Even though floating particles and partial insoluble precipitation are formed, dispersant agent could distribute them in the reacted acid steadily, reducing the possibility of precipitating and plugging (the particular feature also possesses fine rate of metallic corrosion). pH value can be controlled within a certain range by using buffer solution, thus decreasing the amount of secondary precipitation.

The molecular structure of SAV-1 is as shown in Figure 1.

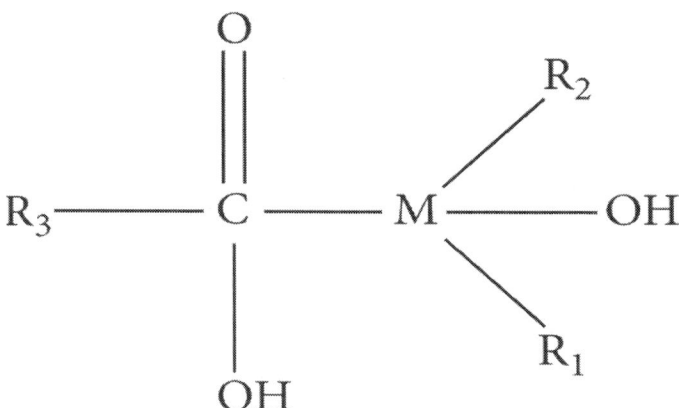

Figure 1: Molecular structure of SAV-1.

The molecular structure of SAV-2 is as shown in Figure 2.

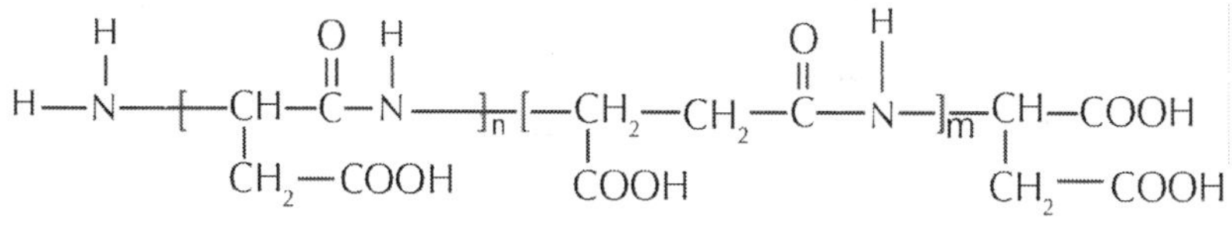

Figure 2: Molecular structure of SAV-2.

EVALUATION ON OVERALL PERFORMANCE OF MULTICHELATING ACID

Retardation Mechanism and Performance Evaluation

Retardation Mechanism

- The main function body of hydrofluoric acid is to unionize HF molecule instead of ionized F^- or HF^{2-}, and surface reaction is the affinity chemisorption of unionized HF molecule and aluminum silicate mineral lattice bond, rather than simply the substitution or generation of hydrogen bond [18, 19].

  The consumption of hydrogen ions is fast, leading to a sharp rise of pH value, which results in hydrofluoric acid affected near borehole area only without further displacement into the formation. SAV-1 and SAV-2 in this claimed system gradually ionize hydrogen ions, and the post production can form a buffer solution to control pH value within a certain range to achieve retardation.

- Regular acid would be mostly consumed on clay surface for it is larger than other minerals', which causes ineffectively utilization of acid as well as destruction of rock framework.

Due to the physical and chemical absorption effects of the multichelating acid, SAV-1 reacts easily with clay and fillers with higher calcium, iron, aluminum component, forming a thin film on clay surface, which can not only promote deeper acidizing but also maintain the integrity of rock framework. The thin film is less than 1 μm under SEM observation, with low solubility in weak acid and water but high dissolving rate in HCl, and it can retard the rapid reaction of clay and acid.

(Montmorillonite)   (Reacted with chelating acid)   (Reacted with mud acid)

Figure 3: SEM of montmorillonite after the reaction with acid.

Figure 3 is the SEM comparison between montmorillonites before and after pictures with mud acid and multichelating acid. The results indicate that when reacted with multichelating acid for 5 minutes, a thin film was formed on the clay surface, which phenomena had not been seen from the mud acid experiment with montmorillonites.

- The addition organic phosphonic acid is an anionic phosphonate with a prominent feature of strong water-wettability. This feature may catalyze the reaction between fluorinated acid and quartz, making the initial low dissolving rate of quartz increased high as time extends, which is helpful to radial permeability improvement of deep formation. Researches demonstrate that this above acid treated clay mineral is agglomerated in toluene and easily dispersed in methanol, which indicates that the water-wet surface will not have bad effect on the reservoir and production of oil and gas development.

Evaluation on Retardation Performance

Mud acid (9%HCl + 1.5%HF) and multichelating acid (5%H_3PO_4 + 4%SAV-1 + 1%SAV-2 + 1.5% NH_4F) were used to react with sandstone minerals, whose composition was 35% quartz, 14% potassium feldspar, 13% albite, 6% dolomite, 2% kaolinite, 9.5% chlorite, 1% Illite, and 1.5% montmorillonite. Experiment temperature was 95°C, with 1 g/10 mL solid-liquid ratio. The experiment results showed that the multichelating acid dissolving rate was far lower than that of mud acid in the early stage, while the dissolution rate of mud acid slightly changed and that of multichelating acid was gradually increased as time went on. After 4 hours of reaction, the dissolution rate of the multichelating acid was nearly equal to that of mud acid, which indicated the better retardation ability of multichelating acid. Therefore, the good retardation performance of multichelating acid prolongs the acid active reaction time and extends the effective distance of acidizing. The results are shown in Figure 4.

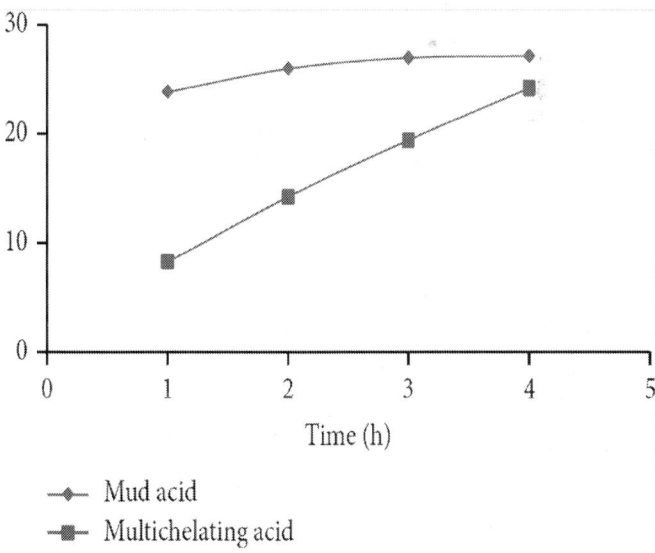

Figure 4: Rock-powder dissolving rate with mud acid and multichelating acid.

Corrosion Inhibition Mechanism and Performance Evaluation

Under high temperature condition, the corrosion rate of acid directly affects the safety of acidizing treatment. The hydroxyl of SAV-1 molecule, where oxygen atom can form coordinate bond with iron ion or partially positive-charged iron atom on the iron surface because of unshared electron pair, thus forms an absorption precipitation layer. The layer covers the iron surface, preventing the diffusion of dissolved oxygen to the metal. Since precipitation membrane is more likely to form in alkaline environment, it inhibits the electrochemical corrosion of the cathodic reaction for OH^- ion generated faster on local cathode area of the precipitation membrane [20, 21]. Additionally, the extended methyl in SAV-1 to the outside of the membrane prevents the diffusion of dissolved oxygen to the metal and protects the mental from corrosion more effectively.

SAV-2 is a water-soluble polypeptide chain, elongated by peptide bond (-CO-NH-). There are quite many polar groups, including carboxyl, carbon, and amino. The oxygen and nitrogen atoms in peptide contain plenty of lone pair electrons. The oxygen atoms in peptide involve ϖ bond, absorbing with d atomic orbital and resulting in directional arrangement of these polar groups on metal surface. However, the nonpolar groups produce great steric hindrance and impede the reduction reaction on cathode by diffusion of O_2 and hence restrict the corrosion of anodic metal. Besides, SAV-2 has a strong chelating ability over many metal ions including ferric ion, decreasing the corrosion caused by depolarization reaction.

Table 1 lists the corrosion rate of steel, without corrosion inhibition agent, of multichelating acid and regular mud acid. The result shows that corrosion rate of mud acid is approximately 10 times of that with multichelating acid, which demonstrating the latter has better corrosion inhibition ability, decreasing the risk of treatment in high-temperature reservoir and lowering the demand for corrosion inhibition agent.

Table 1: Comparison of corrosion rate of multichelating acid and conventional mud acid

	9% HCl + 1.5% HF	5% H_3PO_4 + 4% SAV-1 + 1% SAV-2 + 1.5% NH_4F
Before corrosion		
After corrosion		
Corrosion rate (g/m²·h)	586.75	53.92
Conditions	Temperature: 120°C; time: 4 h	

Scale Removing Mechanism and Performance Evaluation

Sulfate Scales Inhibiting/Removing Mechanism and Performance Evaluation of Multichelating Acid System

Carbonate scale can be easily removed by HCl, leaving sulfate scale to be the key technical problem. There are three steps during the growth and sedimentation of $CaSO_4$ crystal, the process of forming oversaturated solutions: the generating of crystal nucleus, the growing of crystal nucleus, and the forming of crystal. Once any step is broken, the forming of scale would be inhibited or slowed down. The function of inhibitor is to control one or several steps to prevent precipitation. Compared with other acids, multichelating acid has advantage in precipitation inhibiting and removing ability.

(1) Lattice Distortion. In the growth of crystal, there exists vacancy, dislocation, and other lattice defect, or embedded structure distortion with the changing of external condition [22]. Each crystal

plane develops unevenly in a single lattice. The spatial difference in crystal makes the crystal unstable and easy to break when the environment changes. The external reason can be mechanical failure and variation of oversaturation, but most importantly the change of chemical environment. $CaSO_4$ contains ionic lattice, whose growth follows strict sequence. Only when a positive Ca^{2+} collides with negative SO_4^{2-} from another atom, the combination could happen. Therefore, $CaSO_4$ scale is hard scale with certain direction and strictly sequential arrangement. Scale-inhibitor is absorbed on crystal, adulterating inside cell lattice, occupying the active point in polymer, which leads to a more serious distortion in the crystal further growth. Or inner stress raises, and large and irregular amorphous particle is formed, thus making the crystal easy to break, and growth is obstructed.

By comparing the SEM photos before and after the addition of SAV-1/SAV-2, an apparent shape change in crystal can be observed. Before adding SAV-1/SAV-2, the $CaSO_4$ crystal had sharp edges in regular form with smooth surface, where thin bands of crystal were arranged in a tight and ordered way (as shown in Figure 5). After adding SAV-1/SAV-2, the crystal form changed significantly, which had irregular form with shorter length, rough surface, and loose and disorderly arrangement; in addition, minor cracks appeared. The crystal experienced a serious distortion (as shown in Figure 6).

200x   700x   1000x

Figure 5: Plank and pillar-like gypsum crystal before SAV-1/SAV-2 treatment. Annotation: 200x, 700x, and 1000x are the magnification under observation.

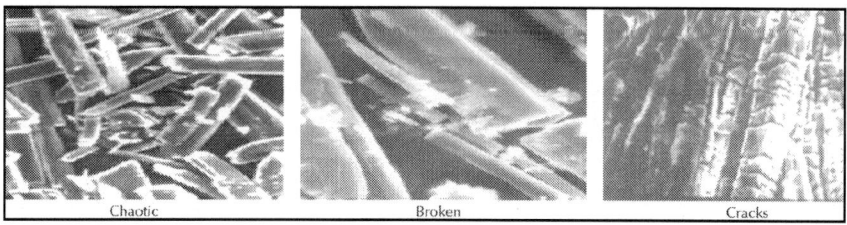

| Chaotic | Broken | Cracks |

Figure 6: Plank and pillar-like gypsum crystal after SAV-1/SAV-2 treatment.

(2) Chelating Solubility. Crystallization is a phase transition process developed in the system microdomain. Besides electrostatic force, oversaturation of crystal material exists as reaction force. Crystallization can be divided into two steps: generation and growth of crystal nucleus. And the addition of inhibitor affects both processes.

SAV-2 structure formula mainly contains -COOH and -NHCO- functional groups, a liner polymeric scale inhibitor with concentration and anion characteristics. SAV-2 can form a stable chelate with Ca^{2+} in water, thus lowering the oversaturation level of $CaSO_4$ and inhibiting formation of scale. Figure 7 is the sketch of this reaction.

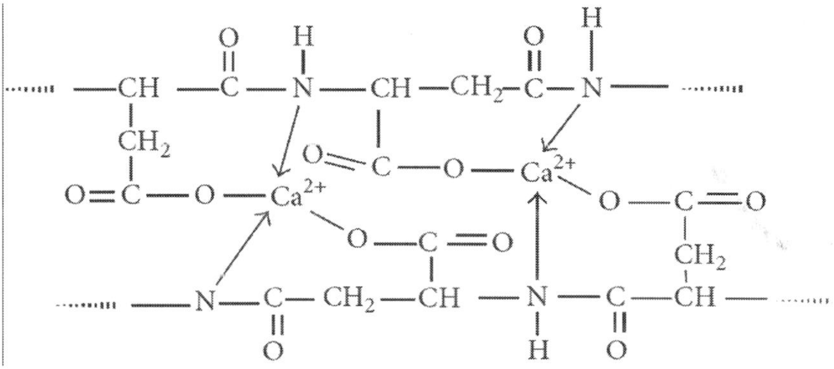

Figure 7: Chelating reaction of SAV-2 and Ca^{2+}.

SAV-1 dissociates H^+ and negatively charged phosphate group, $-PO(OH)_2$, from water, and the latter group provides coordinate electron for Ca^{2+} on $CaSO_4$ surface lattice, constituting chelate with a hexatomic ring. The chelate product has better solubility than that of calcium and magnesium salt; thus calcium and magnesium ions are stabilized in high concentrated chelate-bearing water, inhibiting scale deposition. The hexatomic ring structure formed by SAV-1 and Ca^{2+} is shown in Figure 8.

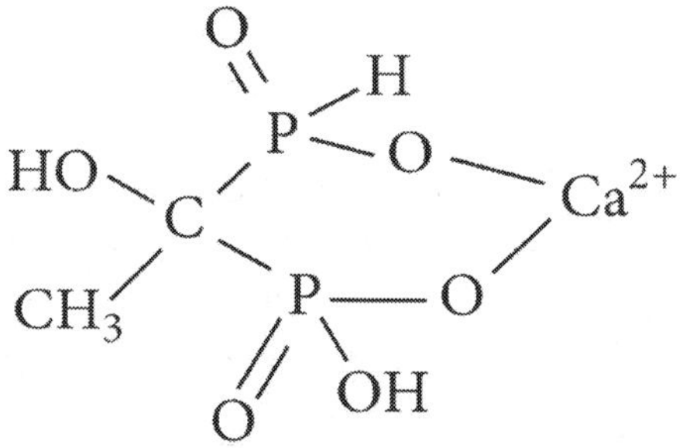

Figure 8: Chelating reaction of SAV-1 and Ca^{2+}.

Also, the acid radical negative anion can react with the Ca^{2+} in formed crystals, making $CaSO_4$ microcrystals hard to arrange in strict lattice orders during collision process; therefore large crystal is uneasy to form. Since solubility is improved for the remaining small grain range of crystals, the multichelating acid also inhibits CaF_2 secondary precipitation effectively based on the same mechanism.

(3) Electrostatic Interaction. SAV-1/SAV-1 dissociates polyaspartic ion, organic phosphonic ion, and hydrogen ion from water, the nitrogen atoms in these acid radical anion and molecule, plus the oxygen atom in carboxyl and phosphine group, resulting in gathered potential scaling microcrystals and repulsed to each other due to the same negative charge (as shown in Figure 9). Therefore, the collision among, the formation of large crystals, the conduction between micro-crystals and metal transmission surface and the formation and growth of scale have all been hindered.

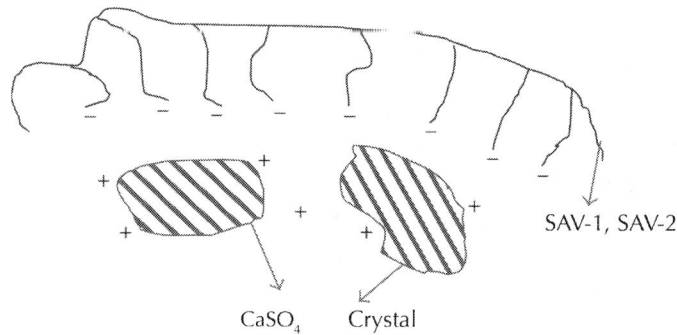

Figure 9: Intercrystalline electrostatic repelling effect.

X-ray diffusion (XRD) is another effective way to study the growth of crystal. When electron rays beams illuminate on a sample, the components trigger multiple X-rays with corresponding features. Among these emitted rays, angles exist between the X-rays and crystal surface satisfying the Prague diffraction condition. The comparison of the diffraction intensity, variation of diffraction angle, crystal axis and other parameters of the scaling samples' X-ray diffraction diagrams, indicating the variations in the degree of fragment, degree of distortion and crystal system with or without scaling inhibitor [23, 24]. Figures 10 and 11 show the XRD diagrams before and after adding SAV-1/SAV-2 into the $CaSO_4$ scale.

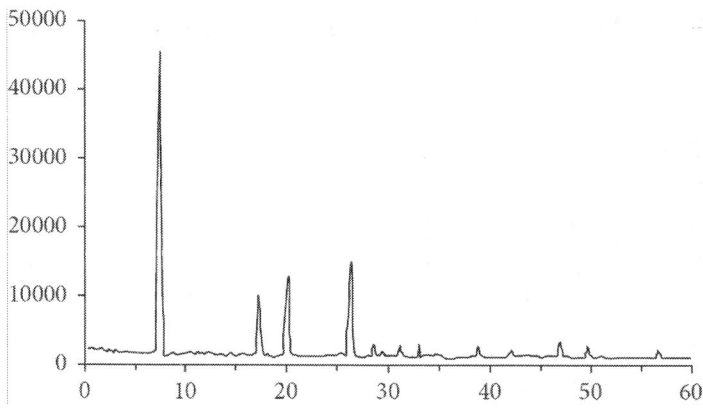

Figure 10: XRD diagram before adding SAV-1/SAV-2.

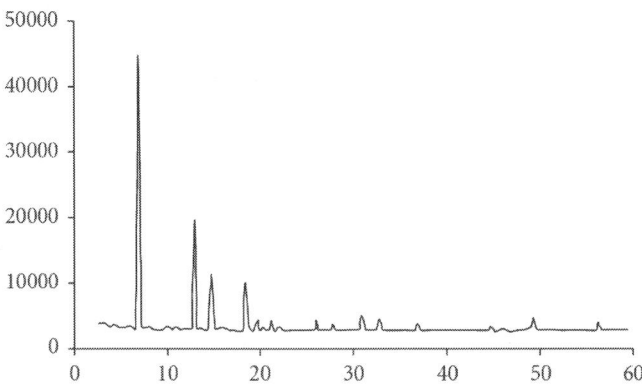

Figure 11: XRD diagram after adding SAV-1/SAV-2.

Generally, there are three forms of $CaSO_4$ scale: gypsum (calcium sulfate carrying two crystalline water particles), calcium sulfate hemihydrate, and anhydrous calcium sulfate. Comparing Figures 10 and 11 with standard XRD diagram, it is certain that crystal type of both samples is gypsum, which belongs to monoclinic system.

It can be spotted from the figures that the diffraction angles before and after adding SAV-1/SAV-2 are almost the same (shown in Table 2), but at some certain diffraction angles the X-rays' relative intensity has changed. After adding SAV-1/SAV-2, the unit cell constant has ascended compared with the straight calcium sulfate (shown in Table 3). The enlargement of unit cell size also demonstrates the entre of SAV-1/SAV-2 or absorption on the crystal surface in crystallization of calcium sulfate, which makes the unit cells of calcium sulfate accumulate in an irregular way; thus different forms are observed.

Table 2: Comparisons of diffraction data of $CaSO_4 \cdot 2H_2O$

	Peak number	$(10^{-10}\,m)$	2θ	h	l	k
Straight crystal	1	7.58121	11.6731	0	2	0
	2	4.2811	20.7488	1	2	1
	4	3.06593	29.127	0	0	2

After treatment of SAV-1/SAV-2	1		11.6/62	0	2	0
7.61214						
	2	4.32105	20.7504	1	2	1
	4	3.13434	29.1294	0	0	2

Table 3: Unit cell constant of $CaSO_4 \cdot 2H_2O$

Samples	$(10^{-10}\,m)$	$(10^{-10}\,m)$	$(10^{-10}\,m)$
Standard	5.68	15.18	6.51
Straight	5.67	15.13	6.51
After treatment of SAV-1/SAV-2	5.71	15.27	6.54

Based on crystallization spiral dislocation theory [25], the points of increasing activity are limited, and when dislocation occurs after active points react with scale-inhibited agent, continuous growth is obstructed. Hence, even a low concentration of scale-inhibiting agent can inhibit the generating of crystal, thus inhibiting the formation of $CaSO_4$.

Silica Gel Precipitation Inhibition Mechanism and Performance Evaluation

Mechanism of Inhibiting Silica Gel Precipitation

- Effects on Surface Properties of Silicon Polymers. Due to physical or chemical effects, scale inhibiting dispersing agent after ionization into anionic will obtain strong absorptive capability. The anions are absorbed on the surface of silicon polymers, changing the physical and chemical properties of silicon polymers' surface, making them acquire same negative charges and causing mutual electrostatic repulsion between particles. These reactions avoid accumulated growth of silicon polymers, suspend disperse particles in water, and block the silicon polymerization from single to dimer or multimer [26]. Meanwhile, with the negatively charged inhibiting dispersant scale agent absorbed on silicon polymers surface, silicon polymers are unlikely to form silicate scales by losing water with the enhanced hydrophilic.

- Dispersion Effects on Silicon Polymers. When an anion of scale inhibitor dispersing agent absorbs one or more silicon polymers, same electronic charge as the anion is with electrostatic repulsion, which prevents collision and the formation of polymers. It is known as flocculation of polymer anion and crystalloid particles, which makes crystalloid be absorbed on polymer chains and also gathers the potential scaling crystalloids to a certain extent [27]. When the products of absorption meet other copolymer molecules or diffuse to a relatively high concentration of polymers area, they offer the absorbed particles to other polymer molecules and finally form a state of balance. This is the dispersion effect on flocculated crystalloids by polymer anion. By such effect of flocculation and dispersion, silicon polymers could keep suspension in aqueous solution stably.

Evaluation of Scale Inhibition Capability

Using $Na_2SiO_3 \cdot 9H_2O$ to prepare 2 kinds of solution during the experiments as comparison (Solution 1, the concentration of silicon ion is 4000 mg/L, containing 1.5 mL SAV-1 and 0.5 mL SAV-2; Solution 2, the concentration of silicon ion is 2000 mg/L without chelating agent), volumes of both solutions are 50 mL, with 2.5 pH value. These two kinds of solutions were heated in water bath under 95°C; after cooling down and filtering, molybdenum blue photometric method was used to measure the content of silicon ion and weight method was used to measure the amount of silica gel precipitate. The results are shown in Table 4.

Table 4: Chelating agent's effect on silica gel precipitate

Sample	30 min		90 min		120 min	
	Concentration of Si (mg/L)	Amount of silica gel precipitate (mg)	Concentration of Si (mg/L)	Amount of silica gel precipitate (mg)	Concentration of Si (mg/L)	Amount of silica gel precipitate (mg)
No chelating agent (2000 mg/L)	1243.12	129.48	765.29	211.22	586.44	241.82
Chelating agent added (4000 mg/L)	3810.63	32.39	3613.13	66.18	3020.63	167.54

The results in Table 4 demonstrate the great inhibiting capability of chelating agent of the formation of silica gel precipitate. Moreover, by comparing the standard silicon ion solution before and after chelating agent addition (as shown in Figure 12), observation shows that the sample precipitates a lot after 30 minutes of heating without chelating agent, and the amount of precipitation increases with the heating time. However, though the sample with chelating agent has a high concentration of silicon ion, the solution still remains transparent after 90 minutes of heating. Precipitate appears after heating for 120 minutes and becomes turbid.

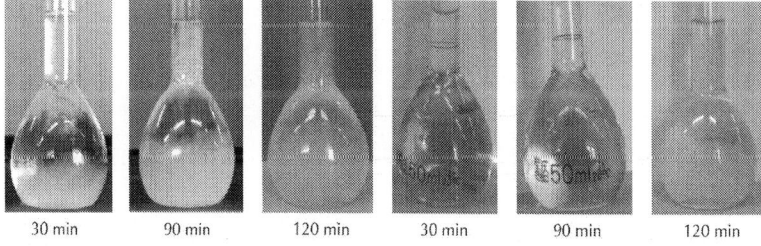

| 30 min | 90 min | 120 min | 30 min | 90 min | 120 min |

Figure 12: Comparison of chelating agent added before and after in Si ion standard solution.

Mechanism of Inhibiting Fluosilicate Precipitation

When sandstone reservoir is treated by HF acid system or other acid systems where HF is produced during the process, the primary reaction between HF and aluminosilicate produces fluosilicic acid, whose reaction with Na^+, K^+, Ca^{2+} in formation water caused reservoir damage by generated silicofluoride precipitation. The mechanism of silicofluoride precipitation inhibition by multichelating agent is basically the same as that of silica gel precipitation inhibition. Generally speaking, organic phosphonic acid has better absorption capability than polycarboxylate acid on crystalloid surface, with a relatively poor dispersion capability. Therefore, by combining chelating agent and chelating dispersant, the inhibition effect on potassium (or sodium) fluorosilicate crystallization can be enhanced, improving scale inhibiting effect [28, 29].

To evaluate the effect of acid on silicofluoride precipitation inhibition, scale inhibiting effect evaluation experiments were conducted [30].

Four kinds of solutions were prepared for comparison experiments: Solution 1: concentration of Na^+ is 0.25 mol/L (or concentration of K^+ is 0.25 mol/L), and concentration of fluosilicic acid is 0.25 mol/L, no scale inhibitor contained. Solution 2: concentration of Na^+ is 0.25 mol/L (or concentration of K^+ is 0.25 mol/L), and concentration of fluosilicic acid is 0.25 mol/L, containing 12.5 mL inhibitor SAV-1 (10 mg/mL). Solution 3: concentration of Na^+ is 0.25 mol/L (or concentration of K^+ is 0.25 mol/L), and concentration of fluosilicic acid is 0.25 mol/L, containing 12.5 mL inhibitor SAV-2 (10 mg/mL). Solution 4: concentration of Na^+ is 0.25 mol/L (or concentration of K^+ is 0.25 mol/L), and concentration of fluosilicic acid is 0.25 mol/L, containing 12.5 mL mixed inhibitor of SAV-1 (5 mg/mL) and SAV-2 (5 mg/mL).

Stainless steel pieces were immersed in beakers filled with 1000 mL of the four solutions separately. After 12 hours of heating in water bath at 50°C, those pieces were taken out, dried, and weighted.

Formulation of the scale inhibition ratio is as follows:

$$\eta = \frac{G_1 - G_2}{G_1} \times 100\%.$$

(1)

In this formulation, G_1 is the weight of stainless steel piece without scale inhibitor addition (g); G_2 is the weight of stainless steel piece with scale inhibitor addition (g); η is the scale inhibition ratio (%).

Table 5 indicates that SAV-2 or the mixture of SAV-1 and SAV-2 has relatively strong inhibiting capability to fluosilicate precipitation.

Table 5: Experimental results of scale inhibitor on the inhibition of sodium fluosilicate precipitation

Scale inhibitor	Scale weight (g)	Scale inhibition ratio (%)
Comparing sample	0.1742	—
SAV-1	0.1536	11.81
SAV-2	0.0582	67.57
SAV-1 + SAV-2 (1 : 1)	0.0632	63.74

CONCLUSION

- The concentration of unionized HF could be reduced to achieve retardation by decreasing the concentration of H^+ in acid; the use of synergic method of chelation and dispersion could prevent/reduce the generation of secondary precipitation; pH value controlled at a certain range with buffer solution could partially decrease secondary precipitation; therefore a new acid system is devised: H_3PO_4 + SAV-1 + SAV-2 + NH_4F.

- Experiments indicate that chelating acid has good performance on retardation and dissolution capability. H_3PO_4 and SAV-1 can slowly ionize H^+; meanwhile the reactant of SAH and SAV-1 can function as buffer solution, which controls pH value and HF concentration within a certain range to achieve retardation. SAV-1 in acid slows the reaction rate by chemical and physical absorption.

- It has been verified that high temperature multichelating acid shows great corrosion inhibition performance, and a few corrosion inhibitors could meet the requirement of acidizing treatment. SAV-1 is a cathodic corrosion inhibitor, mainly focusing on inhibiting cathodic corrosion, while SAV-2 is able to impede the anodic corrosion, and there exists associated corrosion inhibition effect of these two.

- The chelating agents SAV-1 and chelating dispersant SAV-2 of high temperature multichelating acid could eliminate sulfate scale, whose mechanisms are lattice distortion, chelating solubilization, and electrostatic interaction.

- Multichelating acid effectively inhibits the formation of secondary precipitation of fluosilicate, silica gel, and so forth, whose mechanisms are the surface properties' change and dispersion effect on silicon polymers by the effect of SAV-2 and SAV-1. Since the chelating solubilization of multichelating acid system can inhibit the precipitation of CaF_2, potential secondary damage can be eliminated effectively during acid treatment.

ACKNOWLEDGMENTS

The authors wish to thank the management of Southwest Petroleum University and Petro China Changqing Oilfield Company, for their permission to publish this paper and their assistance in applying these new techniques. Thanks are also to Lu Zhang and Zhi Li for their help with paper.

REFERENCES

1. E. P. da Motta, B. Plavnik, R. S. Schechter, and A. D. Hill, "Accounting for silica precipitation in the design of sandstone acidizing," SPE Production and Facilities, vol. 8, no. 2, pp. 138– 144, 1993.

2. E. C. Shuchart and R. D. Gdanski, "Reducing aluminum compound precipitation following subterranean formation acidizing," U.S. Patent no. 6,531,427, 2003.

3. R. L. Thomas, H. A. Nasr-El-Din, J. D. Lynn, and S. Mehta, "Precipitation during the acidizing of a HT/HP illitic sandstone reservoir in eastern saudi arabia: a laboratory study," in Proceedings of the SPE Annual Technical Conference and Exhibition, Society of Petroleum Engineers, pp. 3239–3254, October 2001.

4. S. A. Ali, C. W. Pardo, Z. Xiao et al., "Effective stimulation of high-temperature sandstone formations in East Venezuela with a new sandstone acidizing system," in Proceedings of the SPE International Symposium and Exhibition on Formation Damage Control, Society of Petroleum Engineers, 2006.

5. N. Kume and R. van Melsen, "New HF acid system improves sandstone matrix acidizing success ratio by 400% over conventional mud acid system in Niger Delta Basin," in Proceedings of the SPE Annual Technical Conference and Exhibition, SPE 56527, 1999.

6. M. W. Bodine Jr. and T. H. Fernalld, "Edta dissolution of gypsum, anhydrite, and Ca-Mg carbonates," Journal of Sedimentary Research, vol. 43, no. 4, pp. 1152–1156, 1973.

7. T. N. Tyler, R. R. Metzger, and L. R. Twyford, "Analysis and treatment of formation damage at Prudhoe Bay, Alaska," Journal Of Petroleum Technology, vol. 37, no. 7, pp. 1010–1018, 1985.

8. M. A. Sayed, H. A. Nasr-Ei-din, and C. A. de Wolf, "Emulsified chelating agent: evaluation of an innovative technique for high temperature stimulation treatments," in Proceedings of the

SPE European Formation Damage Conference and Exhibition, pp. 385–404, Society of Petroleum Engineers, June 2013.

9.   B. G. Al-Harbi, N. M. Al-Dahlan, and M. H. Al-Khaldi, "Evaluation of chelating-hydrofluoric systems," in Proceedings of the International Petroleum Technology Conference (IPTC '13), 2013.

10. G. D. Dean, C. A. Nelson, S. Metcalf, R. Harris, and T. Barber, "New acid system minimizes post acid stimulation decline rate in the Wilmington Field, Los Angeles County, California," in Proceedings of the SPE Western Regional Meeting, SPE-46201- MS, Bakersfield, Calif, USA, May 1998.

11. M. A. Mahmoud, A. H. Nasr-El-Din, C. de Wolf, J. N. LePage, and J. H. Bemelaar, "Evaluation of a new environmentally friendly chelating agent for high-temperature applications," in Proceedings of the International Symposium and Exhibiton on Formation Damage Control. Society of Petroleum Engineers (SPE '10), Lafayette, La, USA, February 2010.

12. J. N. LePage, C. A. de Wolf, J. H. Bemelaar, and H. A. Nasr-ElDin, "An environmentally friendly stimulation fluid for hightemperature applications," SPE Journal, vol. 16, no. 1, pp. 104– 110, 2011.

13. W. Frenier, M. Rainey, D. Wilson, D. Crump, and L. Jones, "A biodegradable chelating agent is developed for stimulation of oil and gas formations," in Proceedings of the SPE/EPA/DOE Exploration and Production Environmental Conference, SPE-80597-MS, Society of Petroleum Engineers, San Antonio, Tex, USA, March 2003.

14. W. Frenier, D. Wilson, D. Crump et al., "Use of highly acidsoluble chelating agents in well stimulation services," in Proceedings of the SPE Annual Technical Conference and Exhibition, Society of Petroleum Engineers, 2000.

15. M. A. Mahmoud, H. A. Nasr-El-Din, C. A. de Wolf, and A. K. Alex, "Sandstone acidizing using a new class of chelating agents," in Proceedings of the International Symposium on Oilfield Chemistry, Society of Petroleum Engineers, pp. 1–17, April 2011.

16. A. H. A. Ali,W.W. Frenier, Z. Xiao, and M. Ziauddin, "Chelating agent-based fluids for optimal stimulation of high-temperature wells," in Proceedings of the SPE Annual Technical Conference and Exhibition, Society of Petroleum Engineers, 2002.

17. S. A. Ali, E. Ermel, J. Clarke, M. J. Fuller, Z. Xiao, and B. P. Malone, "Stimulation of high-temperature sandstone formations from West Africa with chelating agent-based fluids," SPE Production and Operations, vol. 23, no. 1, pp. 32–38, 2008.

18. L.-M. Zhang, "Type for hydrofluoric acid dissolution effect of montmorillonite clay," Drilling Fluid and Completion Fluid, vol. 10, no. 5, pp. 28–32, 1993.

19. L.-M. Zhang, "Effect of hydrochloric acid in mud acid through sandstone acidizing," Oil Drilling & Production Technology, vol. 16, no. 6, pp. 51–57, 1994.

20. S.-G. Zhang, Theoretical study on the operation mechanism of scale and corrosion inhibitors Ph.D. thesis., 2006.

21. C.-W. Cui, S.-F. Li, H. Yang et al., "Study of corrosion inhibitors PBTCA, HEDP and ATMP," Materials Science & Technology, vol. 14, no. 6, pp. 605–611, 2006.

22. Y. Q. Zheng, E. W. Shi, W. J. Li, B. G. Wang, and X. F. Hu, "Research and development of the theories of crystal growth," Journal of Inorganic Materials, vol. 14, no. 3, pp. 321–331, 1999.

23. Y.-X. Yang, X-Ray Diffraction Analysis, Shanghai Jiao Tong University Press, 1989.

24. T.-S. Li, Foundation of Crystal X-ray Diffraction Study, Metallurgical Industry Press, 1990.

25. N. M. Min, Physical Foundation of Crystal Growth, Shanghai Science Technology Press, 1982.

26. J. Li and Z.-Y. Lin, "A Study on Scale Inhibiting Mechanism of HEDP Based on Molecule Simulation," Journal of Tongji University (Science and Technology), vol. 34, no. 4, pp. 518–522, 2006.

27.   Y.-X. Cao, J. Yang, and J. Li, "Study of inhibition HEDP and PBTCA," Journal of Tongji University, vol. 32, no. 4, pp. 556–560, 2004.

28.   L.-J. Yang, Y.-X. Zhang, and Y.-H. Huang, "Effect of scale inhibitor-dispersant on the crystallization of potassium (sodium) fluosilicate," Chemical Industry and Engineering, vol. 19, no. 1, pp. 1–5, 2002.

29.   Y. Lin-jun and Z. Yun-xiang, "Kinetics of growth of potassium (Sodium) fluosilicate crystal," Chemical Engineering, vol. 30, no. 2, pp. 15–18, 2002.

30.   G.-N. Feng, Y.-E. Chen, H.-L. Zhu, Y.-Q. Li, R. Zu, and H.-Q. Tang, "An improvement in complexometic titration for evaluating the efficiency of scale inhibitors," Oilfield Chemistry, vol. 21, no. 3, pp. 284–286, 2004.

Microbial Hydrocarbon Degradation: Efforts to Understand Biodegradation in Petroleum Reservoirs

Isabel Natalia Sierra-Garcia[1] and
Valéria Maia de Oliveira[1]

[1]Microbial Resources Division, Research Center for Chemistry, Biology and Agriculture (CPQBA), University of Campinas, Campinas, Sao Paulo, Brazil

INTRODUCTION

The understanding of the phylogenetic diversity, metabolic capabilities, ecological roles, and community dynamics taking place in oil reservoir microbial communities is far from complete.

The interest in studying microbial diversity and metabolism in petroleum reservoirs lies mainly but not only on providing a better comprehension of biodegradation of crude oils, since it represents a worldwide problem for petroleum industry. Generally, biodegradation of oil affects physical and chemical properties of the petroleum, resulting in a decrease of its hydrocarbon content and an increase in oil density, sulphur content, acidity and viscosity, leading to a negative economic consequence for oil production and refining operations [1,2]. Another important point for studying biodegradation lies on its important role in the global carbon cycle and the direct impact on bioremediation of polluted ecosystems. Furthermore, many of the enzymes involved in the degradation pathways are considered key catalysts in industrial biotechnology [3]. Despite these motivations and long recognition of petroleum as a the most important "primary energy" source, at present, microorganisms and factors involved in biodegradation of crude oil hydrocarbons in petroleum reservoirs are still not fully understood. The inaccessibility and complex microbiological sampling of petroleum reservoirs as well as the inherent limitations of the traditional culturing methods conventionally employed can explain this fact. Culture-based techniques have traditionally been the primary tools utilized for studying the microbiology of terrestrial and subsurface environments [4], which allowed the recovery and documentation of a large collection of bacteria capable of hydrocarbon utilization. Studies of numerous aerobic and anaerobic bacterial isolates have revealed mechanisms, which allow them to degrade specific classes of the highly diverse range of hydrocarbon compounds. Therefore, all we know about the degradation of petroleum compounds has come from studying isolated microorganisms. Here, we provide an overview of what is currently known about the mechanisms of aerobic and anaerobic degradation of hydrocarbons, as a result from biochemical and genomic approaches, we give a perspective of the petroleum microbial diversity unraveled so far, and finally we discuss the common oil reservoir characteristics that can be used to predict the most probable mechanism of degradation into deep petroleum reservoirs.

It is well known that microbial diversity in environment is several orders of magnitude higher than the one assumed based on previous cultivation methods [5]. A particularly large number of novel techniques have been developed, which now allow the determination of the *in situ* microbial diversity and activity on a particular site, screening for a particular gene or activity of interest, gene quantification, and DNA and mRNA sequencing and analysis from total communities. This book chapter will address how the implementation of such culture-independent molecular methods allow the access to the microbial diversity and metabolic potential of microorganisms and bring novel information about microbial diversity and new pathways involved in biodegradation processes taking place in petroleum reservoirs. This information will certainly contribute to a broader perspective of the biodegradation processes and corroborate with previous findings that degradation of pollutants in many cases is carried out by microbial consortia rather than a single species [6], where key species and catabolic genes are often not identical to those that have been isolated and described in the laboratory [7, 8].

MICROBIAL DIVERSITY IN OIL RESERVOIRS

Recognition of indigenous microbiota harbored by oil reservoirs has been discussed for a long time. Actually, determining the nature of isolated microorganisms from oil reservoirs (indigenous or nonindigenous) is a difficult issue concerning petroleum microbiologists. The reasons for this controversy rely mainly on the difficulty of aseptic sampling in deep oil reservoirs. This means that microorganisms observed in oil field fluids conceivably could be contaminants introduced during drilling operations and/or during sample retrieval, or could be material sloughed from biofilms growing in installed pipes. Another reason for skepticism is the commonplace practice of "water- flooding" (injection of surface waters or re-injection of natural formation waters to maintain reservoir pressure for oil production); since in this case microbes would be introduced during injection and therefore would not necessarily represent indigenous species [9].

In addition to this controversy, there is the fact that petroleum reservoirs are considered extreme environments where *in situ* conditions, like high pressure, temperature, salinity and anaerobic conditions, are considered as inhospitable to microbial activity. In fact, perception of deep subsurface as a sterile environment has only changed during the past two decades with the increasing awareness of the ability of microbes to colonize extreme environments. Actually, with the use of more sophisticated and appropriate sampling and cultivation techniques, as well as the application of molecular biological techniques to oil field fluids, the dogma of the sterile deep subsurface has been dispelled [9]. Rather, it has become clear that many oil reservoirs do harbor indigenous microbes (*e.g.* the genera *Geotoga* and*Petrotoga* isolated only from oil reservoirs) [10]. Nowadays it is clear that worldwide petroleum reserves are dominated by deposits that have been microbially degraded over geological time and biodegraded petroleum reservoirs represent the most dramatic manifestation of the deep biosphere [11].

In spite of the polemics on which micro-organisms would actually be native and which would be contaminants in oil reservoirs, a wide range of microbial taxonomic groups have been identified in oil reservoirs geographically distant using traditional techniques adapted to *in situ* conditions, as described by L'Haridon et al. [12], Grassia et al. [13] and reviewed by Magot et al [14], or combined with cultivation-independent molecular methods, as reported by Orphan et al. [15]. Table 1 summarizes the various physiological and taxonomical groups and species that have been isolated from oil reservoirs.

ASPECTS FROM OIL RESERVOIR DETERMINING MICROBIAL DEGRADATION

For a long time, the mechanism considered to be prevalent for oil degradation in petroleum reservoirs was the well documented aerobic microbial metabolism and it has long been thought that the flow of oxygen through meteoric waters was necessary for in-reservoir petroleum biodegradation [16]. This mechanism has been widely accepted despite the fact that oxygen would likely be consumed by oxidation of organic matter in near surface sediments and therefore, would be very unlikely for oxygen to reach deep petroleum reservoirs [11].

Recently, the discovery of the ability of microorganisms to degrade anaerobically hydrocarbon oil components and the detection of metabolites characteristic of anaerobic hydrocarbon degradation in oil samples from biodegraded reservoirs, but not in non-degraded reservoirs or aerobically degraded oils [11], have provided valuable information to determine the processes involved in the degradation of oil reservoirs. Nowadays, evidences of such degradation through anaerobic rather than aerobic processes are becoming more substantial and compelling [17].

It is known that microorganisms in anaerobic conditions can use a variety of final electron acceptors, including nitrate, iron, sulfate, manganese and, more recently, chlorate. Anaerobic degradation has also been coupled to methanogenesis, fermentation and phototrophic metabolism but growth of these microorganisms and, therefore, biodegradation rates are significantly lower compared to aerobic degraders. These anaerobic processes have been demonstrated in surface sediments and pure cultures or enrichments in laboratories [18] and all of them potentially play a role in oil biodegradation in anoxic petroleum reservoirs [11]. However, nitrate, like oxygen, is highly reactive and would likely be completely consumed before it could reach the oil reservoir [17]. In deep reservoirs, the supply of large amounts of Fe (III) or manganese (IV) via meteoric water influx are unlikely due to poor solubility and slow water recharge rates in subterranean cycles. Therefore, iron and manganese, which could be used as electro acceptors for oil oxidation, are unlikely to be responsible for significant compositional changes in the oil, considering their limited availability in the reservoir. Accordingly, oil degradation linked to sulfate reduction and methanogenic would therefore explain the consistent hydrocarbon compositional patterns seen in degraded oils worldwide [17]. Sulfate arises from geological sources, such as evaporitic sediments and limestone, or from the injection of seawater for pressure stabilization, and may lead to significant oil degradation and increased residual-oil sulfur content. Methanogenic oil degradation, on the other hand, does not require external electron acceptors and leads to less overall souring of the oil reservoir. Several studies have described *in vitro* methanogenic degradation of crude oil related compounds [19, 20] Jones et al., 2008), including n-alkanes [21, 20] and aromatic hydrocarbons [17].

Table 1: Summary of bacteria isolated from oil reservoirs worldwide

Organism	Taxonomical group	Metabolism	Origin	Reference
Thermodesul-forhabdus nor-vegicus	Deltaproteobacteria	Sulfate-re-ducer	Oil field in Norway	[22]
Desulfacinum infernum	Deltaproteobacteria	Sulfate-re-ducer	North see petroleum reservoir near Scotland	[23]
Desulfomicrobi-um norvegicum	Deltaproteobacteria	Sulfate re-ducer	Petroleum reservoir in Canada	[24]
Desulfovibrio sp.	Deltaproteobacteria	Sulfate re-ducer	Petroleum reservoir in Canada	[24]

Dethiosulfovibrio peptidovorans	Bacteria, Synergistetes	Sulfate reducer	Oil well in the Emeraude oilfield in Congo, Central Africa,	[25]
Desulfotomaculum thermocisternum	Bacteria, Firmicutes	Sulfate reducer	Oil reservoir in the North sea	[26]
Deferribacter sp.	Bacteria, Deferribacteres	Sulfate reducer	California oil fields	[15]
Halanaerobium congolense	Bacteria, Firmicutes	Thiosulfate- and sulfur-reducing bacterium	African oil field	[27]
Thauera phenylacetica	Betaproteobacteria	Nitrate reducer	Petroleum reservoir in Canada	[24]
Pseudomonas stutzeri	Gammaproteobacteria	Nitrate reducer	Petroleum reservoir in Canada	[24]
Garciella nitratireducens	Bacteria, Firmicutes	Nitrate reducer	Oil field in Tabasco, Gulf of Mexico	[28]
Geobacillus subterraneus, Geobacillus uzenensis	Bacteria, Firmicutes	Nitrate reducer	Petroleum reservoir in China	[29]
Lactosphaera pasteurii	Bacteria, Firmicutes	Fermentative	Petroleum reservoir in Canada	[24]
Propionicimonas paludicola	Bacteria, Firmicutes	Fermentative	Petroleum reservoir in Canada	[24]
Anaerobaculum	Bacteria, Synergistetes	Fermentative	California oil fields	[15]
Thermococcus sp.	Archaea, Euryarchaeota	Fermentative	California oil fields	[15]
Thermococcus sibericus	Archaea, Euryarchaeota	Fermentative	Petroleum reservoir in Western Siberia	[30]

Petrotoga sp.	Bacteria, Thermotogae	Fermentative	California oil fields	[15]
Petrotoga olearia; P. siberica	Bacteria, Thermotogae	Fermentative	Petroleum reservoir in Western Siberia	[12]
Thermoanaerobacter	Bacteria, Firmicutes	Fermentative	California oil fields	[15]
Thermotoga sp.	Bacteria, Thermotogae	Fermentative	California oil fields	[15]
Thermosipho geolei	Bacteria, Thermotogae	Fermentative	Petroleum reservoir in Western Siberia	[12]
Anaerobaculum thermoterrenum	Bacteria, Synergistetes	Fermentative	Oil well in Utah	[23]
Fusibacter paucivorans	Bacteria, Firmicutes	Fermentative	Oil well in the Emeraude oilfield in Congo, Central Africa	[31]
Thermovirga lienii	Bacteria, Synergistetes	Fermentative	Oil reservoir in the North sea	[32]
Methanococcus	Archaea, Euryarchaeota	Methanogen	California oil fields	[15]
Methanococcus thermolithotrophicus	Archaea, Euryarchaeota	Methanogen	North sea old field in Norway	[33]
Methanoculleus	Archaea, Euryarchaeota	Methanogen	California oil fields	[15]
Methanobacterium	Archaea, Euryarchaeota	Methanogen	California oil fields	[15]

Deep subsurface environments such as petroleum reservoirs are logistically much more difficult to study than contaminated shallow subsurface environments [17]. Since in many biodegraded petroleum reservoirs most biodegradation occurs close to the oil water transition zone, it has been proposed that the oil–water transition zone (OWTZ) provides suitable physical and chemical conditions for microbial activity [17].

There are other physical and chemical parameters influencing *in situ* biodegradation. Temperature is one of the main factors which limits oil degradation in reservoir, and, empirically, it has been repeatedly observed that biodegradation does not occur in oil reservoirs with *in situ* temperatures >80-90°C [34]. Salinity is another factor that affects in-reservoir oil biodegradation, especially in combination with temperature [13]. Typically, reservoirs with highly saline waters show limited oil biodegradation [11]. This is consistent with the observations that it has not been possible to cultivate microorganisms from reservoir waters with salinity greater than 100 g/L [13]. Pressure seems to be a less limiting factor, except that it may select for certain physiological types and influences the pH of pore waters by increasing dissolution of CO_2 [9]. The availability of electron donors and acceptors governs the type of bacterial metabolic activities within oil field environments [14]. The potential electron donors include CO_2, hydrocarbons, H_2 and numerous organic molecules. Availability of fixed nitrogen is unlikely to limit microbial activity in reservoirs. However, the availability of water-soluble nutrients, like phosphorus and/ or oxidants (terminal electron acceptors such as ferrous iron, sulfate or CO_2), is more likely to limit *in situ* microbial activity [9]. Nonetheless, physiological characteristics of microorganisms indigenous to petroleum reservoirs shed light on the conditions under which petroleum degradation may occur and the potential degradation mechanisms.

HYDROCARBON DEGRADATION

Hydrocarbons are understood as the compounds that consist exclusively of carbon and hydrogen. Because of the lack of functional groups, hydrocarbons are largely apolar and exhibit low chemical reactivity at room temperature. Differences in their reactivities are primarily determined by the occurrence, type and arrangement of unsaturated bonds. Therefore, in this chapter, we will use the common way to classify hydrocarbons according to their bonding features: i) aliphatic group, which includes straight-chain (n-alkanes), branched-chain and cyclic compounds and ii) aromatic group which includes mono or polycyclic hydrocarbons an many important compounds which also contain aliphatic hydrocarbon chains (e. g., alkylbenzenes).

Already a century ago, bacterial isolates had been reported to use aliphatic and aromatic hydrocarbons as sole carbon and energy sources [35]. Since then, numerous aerobic, and also anaerobic, bacterial isolates have been studied in order to understand the mechanisms which allow them to degrade specific members of the highly diverse aliphatic and aromatic compounds. Degradation by such isolates has been investigated thoroughly and results have revealed that they can completely degrade most classes of hydrocarbons, including alkanes, alkenes, alkynes and aromatic compounds. Such degradation can occur aerobically, with oxygen, or anaerobically, with nitrate, ferric iron, sulfate or other electron acceptors [36].

Efforts to overview the metabolism of hydrocarbons in microorganisms are confronted with the chemical diversity of such compounds and their reactivities, as well as with various microbial life styles [36]. The study of biodegradation is conventionally treated in separate areas: aliphatic vs. aromatic hydrocarbons, aerobic vs. anaerobic degradation pathways, physiology and overall metabolic pathways vs. enzymatic mechanisms and structures, often with limited knowledge and data exchange. Nonetheless, each of these study areas deals with the same central point that is the "metabolic challenge" to guide an apolar, unreactive compound composed only of carbon and hydrogen into the metabolism [36]. The hydrocarbon must be first functionalized and currently it has been recognized that there is a surprisingly diversity of reactions of activation that had evolved in microorganisms (Table 2).

Table 2: Overview of aerobic and anaerobic mechanisms for hydrocarbon activation in bacteria

Mechanisms for hydrocarbon activation		
	Aerobic	**Anaerobic**
Short-Chain non-methane alkanes C2-C10	• Non-heme iron monooxygenase similar to sMMO (C2-C9) • Copper-containing monooxygenase similar to pMMO (C2-C9) • Heme-iron monooxygenases (also refered as soluble Cytochrome P450 (C5-C12)	• Fumarate addition
Long-Chain alkanes >C10	• Heme-Monooxygenase (P450 type) • [Fe2]-Monooxygenase • Non-heme iron monooxygenase (AlkB-related) (C3-C13 or C10-C20) • Flavin-binding monooxygenase (AlmA) (C20- C36) • Thermophilic flavin-dependent monooxygenase (LadA) (C10-C30)	• Fumarate addition • Carboxylation
Aromatic hydrocarbons	• [Fe]-Dioxygenase • [Fe2]-Monooxygenase • [Flavin]-Monooxygenase	• Fumarate addition • Hydroxylation • Carboxylation

Mechanisms for hydrocarbon activation are basically different in aerobic and anaerobic microorganisms. Under oxic conditions, hydrocarbon metabolism is always initiated using molecular oxygen as a co-substrate in mono- or dioxygenase reactions that enable the terminal or sub-terminal hydroxylation of aliphatic alkane chains or the mono or dihydroxylation of aromatic rings [37]. In the hydrocarbon activation under anoxic conditions, some proposed reactions comprise: (1) addition to fumarate by glycyl-radical enzymes, (2) methylation of unsubstituted aromatics, (3) hydroxylation with water by molybdenum cofactor containing enzymes of an alkyl substituent via dehydrogenase, and (4) carboxylation catalyzed by yet- uncharacterized enzymes which may actually represent a combination of reaction (2) followed by reaction (1) [38; 37]. Although all these mechanisms of hydrocarbon anaerobic activation have been proposed, the required signature metabolites and enzymes involved have been characterized only for (1) addition to fumarate (demonstrated for toluene, xylene, ethylbenzene, methylnaphthalene, alkanes and alicyclic alkanes); for (3) hydroxylation (demonstrated for ethylbenzene); and for (4) carboxylation (demonstrated for benzene and naphtalene) [39].

BIOCHEMICAL AND GENETIC PATHWAYS OF MICROBIAL HYDROCARBON DEGRADATION

The enzymatic reactions involved in the aerobic degradation of hydrocarbons by bacteria have been extensively studied for several decades [37]. Genes encoding enzymes for degradation are relatively well understood for aerobic and easily cultivable microorganisms, particularly for a *Pseudomonas*strain, known as *P. putida* GPo1, as well as for the strains *Acinetobacter* sp. ADP1 and *Mycobacterium tuberculosis* H37Rv [39, 40]. On the other hand, the anaerobic hydrocarbon degradation has gained more attention since is supposed to be the predominant mechanism occurring in several polluted environments and oil reservoirs. However, its study is an incipient area because of the peculiarities of the reservoir environment and difficulties that arise from attempts to characterize these communities. Nevertheless, several bacteria from other environments able to use alkanes as carbon source in the absence of oxygen have been described in the last few years [41], but anaerobic bacteria able to degrade hydrocarbons under conditions found in deep petroleum reservoirs have not been isolated so far [2].Figure 1 represents an overview of the main mechanisms and pathways used by microorganisms to degrade hydrocarbon compounds under aerobic and anaerobic conditions.

Aerobic Degradation

Aliphatic Hydrocarbons

In most degradation pathways described, the substrate n-alkane is oxidized to the corresponding alcohol by substrate-specific terminal monooxygenases/hydroxylases. The alcohol is then oxidized to the corresponding aldehyde, and finally converted into a fatty acid. Fatty acids are conjugated to CoA and subsequently processed by – oxidation to generate acetyl-CoA [42, 40]. Subterminal oxidation has also been described for both short and long-chain alkanes [40]. Both terminal and sub-terminal oxidation can coexist in some microorganisms [41]. Initial terminal hydroxylation of n-alkanes in bacteria can be carried out by enzymes belonging to different classes, named: (1) propane monooxygenase (C3), (2) different classes of butane monooxygenase (C2-C9), (3) CYP153 monooxygenases (C5-C12), (4) AlkB-related non-heme iron monooxigenase (C3-C10 or C10-C20), (5) flavin-binding monooxigenase AlmA (C20-C36), (6) flavin-dependent monooxygenase LadA (C10-C30), (7) copper flavin-dependent dioxygenase (C10-C30) [43].

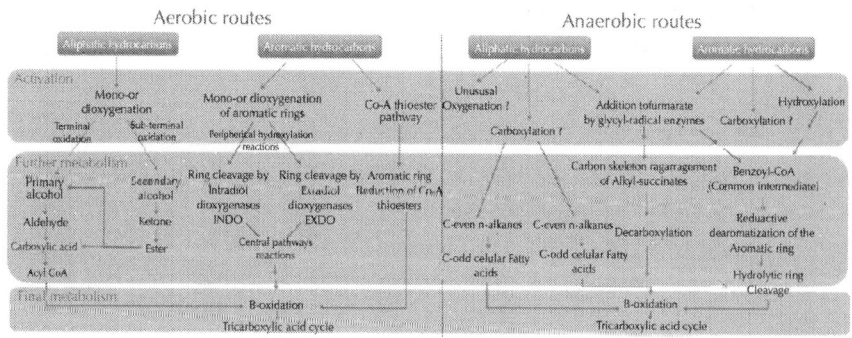

Figure 1: Pathways for aerobic and anaerobic bacterial degradation of hydrocarbon compounds. Two arrows represent more than one reaction.

Among all the alkane activating enzymes, the integral membrane non-heme iron monooxygenase (AlkB) is the best characterized one. Microorganisms degrading medium (C5-C11) and long (>C12)-length alkanes have been frequently related to the presence of *alk*B genes and that is why the presence of such genes have been widely used as functional biomarker for the characterization of aerobic alkane-degrading bacterial populations in several environmental samples [44, 45] and in bioremediation experiments [46, 47]. The degradation pathway of the *alk* system was first described in *Pseudomonas putida* GPo1 (formerly identified as *P. oleovorans* GPo1), where it is located on the OCT plasmid. In this model system, OCT plasmid contains two operons: *alk*BFGHJKL and *alk*ST [48]. The first operon encodes two components of the *alk* system, a particulate non-heme integral membrane alkane monooxygenase (AlkB) and the soluble protein rubredoxin (AlkG), as well as other enzymes involved in further steps. The second operon encodes for a rubredoxin reductase (AlkT and AlkS), which regulates the expression of the *alk*BFGHJKL operon [48, 49]. Since this system was described, AlkB homologous have been found in many alkane-degrading - – and –Proteobacteria and high G + C content Gram-positive bacteria (Actinobacteria) [39] and an increasing collection of alkane hydroxylase gene sequences has allowed the diversity analysis of hydrocarbon-degrading microbial populations in different ecosystems. However, comparisons of cloned *alk*B genes or gene fragments have showed that sequence diversity is very high, even among *alk*B genes within the same species [50].

In despite of the relevance of *alk*B genes as a functional biomarker of alkane-degrading bacterial communities, knowledge on the presence and diversity of *alk*B genes in oil reservoirs is scarce. Tourova et al. [51] analysed *alk*B diversity in thermophilic bacterial strains of the genus *Geobacillus* isolated from oil reservoirs or hot springs. They detected, for the first time, sets of *alk*B gene homologous in thermophilic bacteria, and some strains showed different homologous within the same genome. This fact was explained by the occurrence of horizontal gene transfer among these bacteria. Recently, Li et al. [52] aimed to evaluate *alk*B gene diversity and distribution in production water from 3 oilfields in China through a specific PCR-DGGE method. Results showed that sequences found in the water samples were similar to alkB genes from other corresponding alkane-degrading strains. But at the same time, they showed the presence of a considerable genetic diversity of alkB genes in the wastewater as evidenced by a total of 13 unique DNA bands detected. Studies on the degradation of alkanes in oil reservoirs are currently in a start point, but in the future they certainly will help to understand the process of degradation in oil reservoir.

In comparison to the few efforts in studying alkB system in oil reservoirs, much less is known about the presence of the other enzymatic systems previously listed, which have been described for aerobic degradation of n-alkanes in isolated bacteria or laboratory microcosms. For the most recent elucidated systems for alkane oxidation, named almA and ladA genes, nothing is known about the environmental distribution of these type of genes in petroleum contaminated sites [53] or oil fields, although the LadA complete degradation pathway has been characterized through genome and proteome analysis ofGeobacillus thermodenitrificans NG80-2, a thermophilic strain isolated from a deep oil reservoir in Northern China [54]. Currently, it is believed that there are enzyme systems for alkane degradation which have still not been characterized and that may include new proteins unrelated to those already known [41]. Moreover, in many alkane degraders more than one alkane oxidation system have been observed, which have been reported exhibiting overlapping substrate ranges [39, 40]. These observations point out that in order to characterize and explore metabolic diversity and functions involved in alkane degradation one should take into consideration the high diversity of enzymes capable of initiating such metabolism.

Aromatic Hydrocarbons

The aerobic bacterial catabolism of aromatic compounds involves a wide variety of peripheral pathways that activate structurally diverse substrates into a limited number of common intermediates that are further cleaved and processed by a few central pathways to the central metabolism of the cell [55]. Metabolic pathways and encoding genes responsible for the degradation of specific members of a highly diverse range of aromatic compounds have been characterized for many isolated bacterial strains, predominantly from the Proteobacteria and Actinobacteria phyla [56]. Degradation by such isolates is typically initiated by members of one of the three superfamilies: the Rieske non-heme iron oxygenases (RNHO), the flavoprotein monooxygenases (FPM) and the soluble diiron multicomponent monooxygenases (SDM). Further metabolism is achieved through di- or trihydroxylated aromatic intermediates. Alternatively, activation is mediated by CoA ligases where the formed CoA derivates are subjected to selective hydroxylation [58, 53]. In the case of hydrophobic pollutants, such as benzene, toluene, naphthalene, biphenyl or polycyclic aromatics, aerobic degradation is usually initiated by activation of the aromatic ring through oxygenation reactions catalyzed by RNHO enzymes or, as intensively described for toluene degradation, through members of SDM enzymes [56].

Further intermediates can be catalyzed by two kinds of enzyme, intradiol and extradiol dioxygenases, which represent two classes of phylogenetically unrelated proteins [58]. These enzymes are key enzymes in the degradation of aromatic compounds, and many of such proteins and their encoding sequences have been described, purified and characterized in the last decades [56]. While all intradiol dioxygenases described so far belong to the same superfamily, the extradiol dioxygenases include at least three members of different families. Type I extradiol dioxygenases (e.g. catechol 2,3-dioxygenases and 1,2-dioxygenases) belong to the vicinal oxygen chelate superfamily enzymes. Type II extradiol dioxygenases are related to LigB superfamily (e.g. protocatechuate 4,5-dioxygenases) and the type III enzymes belongs to the cupin superfamily (e.g. gentisate dioxygenases) [53]. However, members of novel superfamilies performing crucial steps in aromatic metabolic pathways are still being discovered [56, 53].

The knowledge of metabolic properties of isolates has allowed the monitoring of the ability of microorganisms to mineralize aromatic hydrocarbons in soils. Typically, these studies have used primers designed based on conserved gene regions and focused on RNHO or SDM as targets for initiating degradation, or on Extradiol dioxygenases (EXDO) cleaving the aromatic ring [59]. These studies range from those searching for a narrow range of genes similar or identical to those observed in type strains using non-degenerated primers to those searching for subfamilies of homologous genes using degenerated primers [59]. However, due to the immense heterogeneity of such enzymes [57], there will never be a pair of primers that will reliably cover the huge diversity of a catabolic gene family in nature [53].

Anaerobic Degradation

Aromatic Hydrocarbons

We have already described the main mechanism for degradation of aromatic compounds in aerobic conditions, where oxygen is not only the final electron acceptor but also co-substrate of two key processes: hydroxylation and cleavage of the aromatic ring by oxygenases. In contrast, in the absence of oxygen, microorganisms use a complete different pathway, based in reductive reactions to attack the aromatic ring [61].

The biochemistry of some anaerobic degradation pathways of aromatic compounds has been studied to some extent; however, the genetic determinants of all these processes and the mechanisms involved in their regulation are much less studied [55]. Recent advances in genome sequencing have led to the complete genetic information for six bacterial strains that are able to anaerobically degrade aromatic compounds using different electron acceptors and that belong to different taxonomic groups of bacteria: denitrifying betaproteobacteria, *Thauera aromatica* and *Azoarcus* sp. EbN1, two alphaproteobacteria, the phototroph *Rhodopseudomonas palustris* strain CGA009 and the denitrifying *Magnetospirillum magneticum* strain AMB-1, and two obligate anaerobic deltaproteobacteria, the iron reducer*Geobacillus metallireducens* GS-15 and the fermenter *Syntrophus aciditrophicus* strain SB [55]. It is worth remembering that, in recent years, important inferences and generalizations have been made about the genetics involved in hydrocarbon metabolism based on these isolated bacteria under conventional laboratory conditions. However, potential novel genes, enzymes and metabolic pathways responsible for degradation processes are probably harbored by yet uncultivated bacteria.

The best understood and apparently the most widespread of these anaerobic mechanisms is the radical-catalyzed addition of fumarate to hydrocarbons, yielding substituted succinate derivatives. This reaction has been recognized for the activation of several alkyl-substituted benzenes as well for n-alkanes [62]. However, understanding of this fumarate-dependent hydrocarbon activation is most advanced in the case of toluene. The key enzyme in this process is the enzyme benzylsuccinate synthase. All enzymes required for -oxidation of benzylsuccinate are encoded by the *bbs* operon. Subsequent degradation of benzoyl-CoA proceeds via reductive dearomatization, hydrolytic ring cleavage, -oxidation to acetyl-CoA units and terminal oxidation to Co_2 [63]. In contrast to the anaerobic metabolism of toluene, degradation of ethylbenzene (and probably other alkylbenzenes with carbon chain of at least 2) is entirely different, despite the chemical and structural similarities between the two compounds, and involves a direct oxidation of the methylene carbon via (S)-1-phenylethanol to acetophenone [55]. Ethylbenzene is anaerobically hydroxylated and dehydrogenated to acetophone, which is then carboxyled and converted to benzoylCoA as the first common intermediate of the two pathways [62].

Genetics of the enzymatic system have been only characterized for these two mechanisms for anaerobic hydrocarbon activation. Genes encoding pathways that involve fumarate addition are typically organized in two operons. One operon includes the three structural genes of the protein catalyzing fumarate addition and the other includes genes required for converting succinate derivates to benzoyl-CoA [64]. Gene sequences and organization are relatively conserved among nitrate-reducing bacteria but differ somewhat from those of the iron reducer *G. metallireducens* [64] and substantially from those of the hexane-degrading nitrate reducer strain HxN1 [65]. Hydrocarbon dehydrogenation pathway is also organized in two operons. One operon contains the structural genes for the first two reactions (ethylbenzene dehydrogenase and 1-phenylethanol dehydrogenase) and the other contains the structural genes for acetophone carboxylase [64].

Kane et al. [66] developed the first real-time polymerase chain reaction (PCR) method to quantify hydrocarbon utilizers based on *bss*A genes of nitrate-reducing Betaproteobacteria. Since then, there have been several additional studies investigating the presence and/or distribution of anaerobic hydrocarbon utilizers in anaerobic environments via functional gene surveys of *bss*A, extending the range of detectable hydrocarbon-degrading microbes to iron and sulfate-reducing Deltaproteobacteria and revealing partially novel, site specific degrader populations [67, 68]. Other *bss*A-based detection studies in impacted environments, as well as studies that combine field metabolomics and molecular tools, are described by other authors [69, 70, 71]. Despite of the role of benzylsuccinate synthase in aromatic hydrocarbon degradation and its use as a biomarker are well documented, there is no study on the presence of this gene in oil reservoirs.

Aliphatic Hydrocarbons

Anaerobic degradation of alkanes has not been extensively studied as for some aromatic compounds. The presumable reasons include the greater attention given to BTEX compounds (benzene, toluene, ethylbenzene and xylenes) because of their classification as priority pollutants [71], also the fact that anaerobic growth with n-alkanes is even slower than that with the alkylbenzenes, and finally the fact that long chain alkanes are poorly soluble and often prevents the cultivation of cells homogeneously in the medium [72]. However, anaerobic degradation of alkanes is equally relevant, since alkanes are quantitatively the most important hydrocarbon components of petroleum, and some are acutely toxic and difficult to remediate [71]. Several anaerobic bacteria capable of degrading n-alkanes with 6 or more carbons in length, particularly hexadecane (C16), using sulfate or nitrate as electron acceptors have been isolated [72, 73].

The two main mechanisms of anaerobic degradation of n-alkanes described involve unprecedented biochemical reactions that differ completely from those employed in aerobic hydrocarbon metabolism [73]. The first involves activation at the subterminal carbon of the alkane by the addition of fumarate, analogously to the formation of benzyl succinate during anaerobic degradation of toluene, however further reactions are completely different involving dehydrogenation and hydration [72]. Studies conducted with established axenic cultures have indicated that anaerobic metabolism of oil allkanes predominantly proceeds via addition of fumarate to the double bound [72]. Although alkylsuccinate metabolites have rarely been detected in oil reservoir fluids [74, 75], they have been reported in oil-contaminated environments as well as in oilfield facilities, where their detection is indicative of in situ microbial degradation of oil alkanes [71, 75]. Alkylsuccinic acids as intermediates of anaerobic alkane oxidation were first studied by Gieg and Suflita [76] when surveying these metabolites in aquifers contaminated with condensate gas, natural gas liquids, gasoline, diesel, alkanes and BTEX. They found alkylsuccinates originating from C3 to C11 alkanes, as well as putative metabolites originating from compounds with one degree of unsaturation, such as alkenes or alicyclic alkanes. Since this report, other studies have detected alkylsuccinate derivates in petroleum contaminated groundwater systems [76], coal beds [70] and oil fields [74, 77]. The formation of alkylsuccinates is catalyzed by a strictly anaerobic glycyl radical enzyme which has been termed as alkylsuccinate synthase or (1-methyl-alkyl)succinate synthase (Ass or Mas). The genes encoding Ass have recently been identified in the alkane degrading sulfidogenic bacteria *D. alkenivoras* AK-01 [78] and *Desulfoglaeba alkanedexens* ALDC[T] [71], as well as in nitrate reducing strains HxN1 [65] and OcN1 [79], all affiliated to the

Proteobacteria phylum [80]. Recently, Callaghan et al. [71] detected *assA* genes in a propane-utilizing mixed culture and in a paraffin-degrading enrichment culture maintained under sulfate-reducing conditions. Despite of no genes for benzyl-and alkylsuccinate synthase were found when environmental metagenome datasets of uncontaminated sites were analyzed in Callaghan et al [71], the authors consider that *assA* gene could be a useful biomarker for anaerobic alkane metabolism.

The second mechanism for alkane anaerobic degradation is the carboxylation, mainly developed from the growth pattern of the sulfate-reducing strain Hxd3 [81], tentatively named as *Desulfococcus oleovorans*. This strain differs from other alkane degraders for converting C-even alkanes into C-odd cellular fatty acids whereas growth on C-odd alkanes resulted in C-even cellular fatty acids [81, 72]. More recently, Callaghan et al. [82] suggested that a carboxylation-like mechanism analogous to the activation strategy previously proposed by So et al. [81] was the probable route for the anaerobic biodegradation of hexadecane in an alkane-degrading, nitrate-reducing consortium. However, in both cases, the hypothetical fatty acid intermediate (2-ethylalkanoate) that should result from the incorporation of inorganic carbon at C-3 of the alkane has never been detected. There is an on-going debate about this initial activation mechanism. From an energetic point of view, the carboxylation of alkanes is not feasible under physiological conditions, unless the concentration of the fatty acid (2-ethylalkanoate) is in the micromolar order of magnitude or less [80].

Other alternative activation mechanisms are proposed for the anaerobic degradation of alkanes. For instance, the mechanism referred as "unusual oxygenation" is used by the strain *Pseudomonas chloritidismutans* AW-1[T], that is assumed to produce its own oxygen via chlorate respiration used for subsequent metabolism of alkanes [60]. Other alternative mechanism considers that activation in the anaerobic methanogenic system may be initiated by an anaerobic hydroxylation reaction [83].

MECHANISMS INVOLVED IN OIL BIODEGRADATION IN PETROLEUM RESERVOIRS

From those microorganisms studied in oilfields, methanogens have received particular attention since they have been isolated and molecularly detected in both low- and high-temperature reservoirs [88, 89]. Their physiological characteristics and potential activity possibly involved in methanogenesis occurring in oil reservoirs have been demonstrated [90]. Furthermore, recently, Jones et al. [20] provided evidence that the patterns of hydrocarbon degradation observed in biodegraded petroleum reservoirs were the result of methanogenic processes. Therefore, microbiological and biogeochemical investigations have indicated that methanogenesis is a widely distributed process in petroleum reservoirs, although still poorly understood [90]. Methanogenesis is the terminal process of biomass degradation. Acetate and hydrogen are the most important immediate precursors for methanogenesis, and are converted into methane by acetoclastic and hydrogenotrophic methanogens, respectively [91]. Acetate can also be a precursor for methanogenesis through syntrophic acetate oxidation coupled to hydrogenotrophic methanogenesis, which is mediated by syntrophic bacteria and methanogenic archaea [92, 93, 94, 95]. Interestingly, acetate is generally abundant in many petroleum reservoirs, at concentrations ranging between 0.3 and 20 mM [96] hence, acetate metabolism is considered an important methane production process in those environments [90].

Cultivation-dependent and -independent approaches have shown the presence of acetoclastic and hydrogenotrophic methanogens and putative syntrophic acetate-oxidizing bacteria in reservoirs [88, 89,102], indicating that there should be two different pathways of acetate metabolism in the environment, namely acetoclastic methanogenesis and syntrophic acetate oxidation coupled with hydrogenotrophic methanogenesis. Some previous studies suggested that syntrophic acetate oxidation was most likely to occur in petroleum reservoirs, based on molecular biological analysis [89] and thermodynamic calculations [98]. In Jones et al. [20], the composition of oil in microcosms exhibiting methanogenic oil degradation is compared to patterns observed in biodegraded oils from the Gullfaks field in the North Sea. Analysis of the methanogenic communities from oil-degrading microcosms revealed a strong selection for CO_2-reducing methanogens against acetoclastic methanogens, and gas isotope modeling also revealed that to match the d13C of methane and carbon dioxide from biodegraded petroleum reservoirs 75–92% of methanogenesis should be via the CO_2 reduction pathway [20, 11].

The reason why syntrophic acetate oxidation predominates over acetoclastic methanogenesis in oil reservoirs remains unclear. There is evidence from studies of oil contaminated aquifers that crude oil can have a detrimental effect on acetoclastic methanogenesis and, in situations where acetoclastic methanogenesis is inhibited, methanogenic alkane degradation via syntrophic acetate oxidation may be thermodynamically the most favorable alternative pathway [11]. Nonetheless, one recent report suggests that acetoclastic methanogenesis may predominate in some methanogenic oil-degrading systems [19]. Although there is currently great interest in how much each of the two pathways contributes to methane production in petroleum reservoirs, no studies are being conducted to address this question [90].

METAGENOMICS AS A TOOL FOR A BETTER COMPREHENSION OF BIODEGRADATION

As stated previously, cultivation-based methods have traditionally been utilized for studying the microbiology in oil fields and have yielded valuable information about microbial interactions and their relations with hydrocarbons [42]. However, nowadays, it is known that only a small fraction of the microbial diversity in nature (1-10%)

can be grown in the laboratory [84, 85, 86]. Therefore, it is assumed that the ecological functions of the majority of microorganisms in nature and their potential applications in biotechnology remain obscure [87].

In metagenomics, total DNA is extracted from appropriately chosen environmental samples, propagated in the laboratory by cloning techniques, submitted to sequence or function-based screenings and/or subjected to large-scale sequence analysis (Fig. 2). Functional screening of metagenomic libraries offer the advantage that it does not rely on sequence homology to known genes, and for this reason, has allowed the isolation of different enzyme classes from several environments. The probability (hit rate) of identifying a certain gene depends on multiple factors that are intrinsically linked to each other: the host–vector system, size of the target gene, its abundance in the source metagenome, the assay method, and the efficiency of heterologous gene expression in a surrogate host [99].

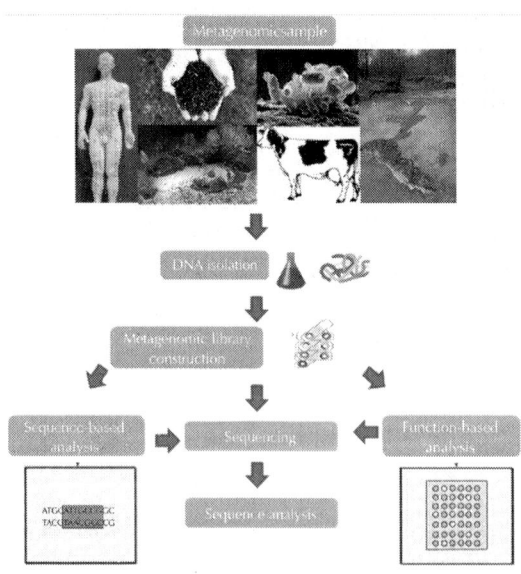

Figure 2: Schematic representation of the different steps for metagenomic analysis.

One of the first studies using metagenomics to study microbial degradation of aromatic compounds was performed by Suenaga and colleagues [100], who constructed a metagenomic library from activated sludge for industrial wastewater. The library was functionally screened for extradiol dioxygenase activities (enzymes for aromatic degradation) and 38 clones were subjected to sequencing analysis [101]. As a result, various types of gene subsets were identified that were not similar to the previously reported pathways performing complete degradation. Moreover, the authors discussed the fact that aromatic compounds in the environment may be degraded through the concerted action of various fragmented pathways. Sierra-Garcia [101] reported the organization of hydrocarbon degradation genes of selected metagenomic fosmid clones derived from a metagenomic library from Brazilian petroleum reservoir and functional screening for hydrocarbon degradation activities. The author found many putative proteins of different aerobic and anaerobic well described catabolic pathways, however the complete catabolic pathways described for hydrocarbon degradation in previous studies were absent in the fosmid clones. Instead, the metagenomic fragments comprised genes belonging to different pathways, showing novel gene arrangements where hydrocarbon compounds were degraded through the concerted actions of these fragmented pathways. These results suggest that there are marked differences between the degradation genes found in microbial communities derived from enrichments of oil reservoir sample and those that have been previously identified in bacteria isolated from contaminated or pristine environments.

However, function-based screening of metagenomic libraries for xenobiotic degradation genes is often considered problematic because of insufficient and biased expression of the heterologous genes in the host *Escherichia coli* [99]. Only a few efforts have been made to solve these problems. In Uchiyama et al. [103], a novel method for function-driven screening is described, which was termed substrate-induced gene expression screening (SIGEX). This high-throughput screening approach employs an operon trap gfp expression vector in combination with fluorescence-activated cell sorting. The screening is based on the fact that catabolic-gene expression is induced mainly by specific substrates and is often controlled by regulatory elements located close to catabolic genes [103]. Using this approach, Uchiyama et al. [103] isolated aromatic-hydrocarbon-induced genes from a metagenomic library derived from groundwater. In Ono et al. [104] another screening strategy was based on functional complementation of a *Pseudomonas putida* host strain containing a naphthalene degrading pathway devoid of the naphthalene dioxygenase (NDO) encoding gene. Two clones were able to restore the ability of the host strain to use naphthalene as a sole carbon source and their genes were similar but no identical to already known operons. The authors refer to the use of other host strains for the construction of metagenomic libraries instead of the well-established *E. coli* as a simpler and economical way to perform function-driven screening in comparison to other reported systems such as SIGEX [103].

In the context of this chapter, several aspects of the hydrocarbon degradation need to be studied to obtain a comprehensive overview of the biodegradation processes that take place in oil reservoirs or petroleum impacted environments. These studies should take into consideration the high diversity of enzymes capable of initiating such metabolism as well as the implementation of integrated studies combining culture and molecular techniques, linking with metabolomics or compound-specific isotope analysis and microcosm studies for a better resolution of in situ microbial activity in petroleum reservoirs.

CONCLUSIONS AND RESEARCH NEEDS

The understanding about biodegraded petroleum reservoirs have advanced considerably in recent years, but the organisms responsible for the *in situ* activity and a quantitative understanding of the factors which control in-reservoir oil biodegradation remain far from complete. The inaccessibility of petroleum reservoirs and inherent difficulties of microbiological sampling from commercially operating oil wells have required a multidisciplinary approach to delineating the study of subsurface petroleum biodegradation, and to date there are still prevailing paradigms relating to hydrocarbon biodegradation processes. This multidisciplinary approach to study *in situ* petroleum degradation should consider molecular biology, microbiology, and geological and geochemical parameters in order to establish the key organisms, biochemical reactions and mechanisms involved in such complex associations. Indeed, the isolation of anaerobic microorganisms capable of utilizing hydrocarbons is essential for a comprehensive understanding of their role and behavior in anoxic habitats and their complex interactions within methanogenic hydrocarbon-degrading communities. In addition, novel approaches, combining functional metagenomics, transcriptomics, metabolomics and other molecular surveys in microcosms are urgently required to better allow access to a more realistic phylogenetic and metabolic diversity governing oil biodegradation in petroleum reservoirs.

REFERENCES

1.   W Roling, 2003The microbiology of hydrocarbon degradation in subsurface petroleum reservoirs: perspectives and prospects. *Res Microbiol*. 154(5), 321-328.

2.   I. M Head, D. M Jones, S. R Larter, 2003Biological activity in the deep subsurface and the origin of heavy oil. *Nature*, 426(6964), 344-52.

3.   W Ismail, and J Gescher, 2012Epoxy coenzyme a thioester pathways for degradation of aromatic compounds. *Appl Fnviron Microbiol*. 7815504351

4.   D. P Chandler, S. M Li, C. M Spadoni, G. R Drake, D. L Balkwill, J. K Fredrickson, F. J Brockman, 1997A molecular comparison of culturable aerobic heterotrophic bacteria and 16S rDNA clones derived from a deep subsurface sediment. *FEMS Microbiol Ecol*. 23131144

5.   M. B Leigh, V. H Pellizari, O Uhlik, R Sutka, J Rodrigues, N. E Ostrom, *et al* 2007Biphenyl-utilizing bacteria and their functional genes in a pine root zone contaminated with polychlorinated biphenyls (PCBs). *ISME J* 1134148

6.   V De Lorenzo, 2008Systems biology approaches to bioremediation. *Curr Opin Biotechnol* 19579589

7.   C Jeon, W Park, P Padmanabhan, C Derito, J Snape, E Madsen, 2003Discovery of a bacterium, with distinctive dioxygenase, that is responsible for*in situ* biodegradation in contaminated sediment. *Proc Natl Acad Sci USA* 1001359113596

8.   R Witzig, H Junca, H. J Hecht, D. H Pieper, 2006Assessment of toluene/biphenyl dioxygenase gene diversity in benzene-polluted soils: links between benzene biodegradation and genes similar to those encoding isopropylbenzene dioxygenases. *Appl Environ Microbiol* 7235043514

9.   J Foght, 2010Microbial comminities in oil shales, biodegraded and heavy oil reservoirs, and bitumen deposits. In: K. N. Timmis (Ed.) *Handbook of Hydrocarbon and Lipid Microbiology.* Berlin, Heidelberg: Springer Berlin Heidelberg.

10.   N. K Birkeland, 2004The microbial diversity of deep subsurface oil reservoirs. *Stud Surface Sci Catal* 151385403

11.   I. M Head, C. M Aitken, N. D Gray, A Sherry, J. J Adams, D. M Jones, A. K Rowan, et al2010Hydrocarbon degradation in petroleum reservoirs. In: K. N. Timmis (Ed.) *Handbook of Hydrocarbon and Lipid Microbiology.* Berlin, Heidelberg: Springer Berlin Heidelberg.

12.   L Haridon, S Reysenbach, A. L Glenat, P Prieur, D Jeanthon, C. (1995Hot subterranean biosphere in a continental oil reservoir. *Nature* 377223224

13.   G. S Grassia, K. M Mclean, P Glenat, J Bauld, A. J Sheehy, 1996A systematic survey for thermophilic fermentative bacteria and archaea in high temperature petroleum reservoirs. *FEMS Microbiol* Ecol 214758

14.   M Magot, B Ollivier, B. K. C Patel, 2000Microbiology of petroleum reservoirs. *Antonie van Leeuwenhoek.* 772103116

15. V. J Orphan, L. T Taylor, D Hafenbradl, and E. F Delong, 2000Culture-dependent and culture-independent characterization of microbial assemblages associated with high-temperature petroleum reservoirs. *Appl Environ Microbiol.* 66270011

16. C. M Aitken, D. M Jones, S. R Larter, 2004Anaerobic hydrocarbon biodegradation in deep subsurface oil reservoirs. *Nature*, 43170062914

17. N. D Gray, A Sherry, C Hubert, J Dolfing, I. M Head, 2010Methanogenic degradation of petroleum hydrocarbons in subsurface environments remediation, heavy oil formation, and energy recovery. *Adv Appl Microbiol.* 7213761

18. F Widdel, R Rabus, 2001Anaerobic biodegradation of saturated and aromatic hydrocarbons. *Curr Opin Biotechnol* 12259276

19. L. M Gieg, K. E Duncan, J. M Suflita, 2008Bioenergy production via microbial conversion of residual oil to natural gas. Appl Environ Microbiol 7430223029

20. D Jones, I Head, N Gray, J Adams, A Rowan, C Aitken, B Bennett, et al2007Crude-oil biodegradation via methanogenesis in subsurface petroleum reservoirs. *Nature*, 451(7175), 176-180.

21. K Zengler, H. H Richnow, R Rossello-mora, W Michaelis, F Widdel, 1999Methane formation from long chain alkanes by anaerobic microorganisms. *Nature* 401266269

22. J Beeder, T Torsvik, and T Lien, 1995*Thermodesulforhabdus norvegicus* gen. nov., sp. nov., a novel thermophilic sulfate-reducing bacterium from oil field water. *Arch. Microbiol* 164331336

23. G. N Rees, G. S Grassia, A. J Sheehy, P. P Dwivedi, B K. C Patel, 1995*Desulfacinum infernum* gen. nov., sp. nov., a thermophilic sulfate-reducing bacterium from a petroleum reservoir. *Int. J. Syst. Bacteriol* 458589

24. [24] A Grabowski, O Nercessian, F Fayolle, D Blanchet, C Jeanthon, 2005Microbial diversity in production waters

of a low-temperature biodegraded oil reservoir. *FEMS microbiology ecology*, 54(3), 427-43.

25. M Magot, G Ravot, X Campaignolle, B Ollivier, B. K Patel, M. L Fardeau, P Thomas, J. L Crolet, J. L Garcia, 1997Dethiosulfovibrio peptidovorans gen. nov., sp. nov., a new anaerobic, slightly halophilic, thiosulfate-reducing bacterium from corroding offshore oil wells. *Int. J. Syst. Bacteriol.* 47818824

26. R. K Nilsen, T Torsvik, T Lien, 1996Desulfotomaculum thermocisternum sp. nov., a sulfate reducer isolated from a hot North Sea oil reservoir. *Int. J. Syst. Bacteriol.* 46397402

27. G Ravot, M Magot, B Ollivier, B. K. C Patel, E Ageron, P. A. D Grimont, P Thomas, J. L Garcia, 1997*Haloanaerobium congolense* sp. nov., an anaerobic, moderately halophilic, thiosulfate- and sulfur-reducing bacterium from an African oil field. *FEMS Microbiol. Lett.* 1478188

28. E Miranda-tello, M. L Fardeau, L Fernandez, F Ramirez, J. L Cayol, P Thomas, J. L Garcia, B Ollivier, 2003*Desulfovibrio capillatus* sp. nov., a novel sulfatereducing bacterium isolated from an oil field separator located in the Gulf of Mexico. *Anaerobe* 997103

29. T. N Nazina, T. P Tourova, A. B Poltaraus, E. V Novikova, A. A Grigoryan, A. E Ivanova, et al2001Taxonomic study of aerobic thermophilic bacilli: Descriptions of *Geobacillus subterraneus* gen. nov., sp. nov. and *Geobacillus uzenensis* sp. nov. from petroleum reservoirs and transfer of *Bacillus stearothermophilus, Bacillus hermocatenulatus, Bacillus thermoleovorans, Bacillus kaustophilus, Bacillus thermoglucosidasius* and *Bacillus thermodenitrificans* to *Geobacillus* as the new combinations *G. stearothermophilus, G. thermocatenulatus, G. thermoleovorans, G. kaustophilus, G. thermoglucosidasius* and *G. thermodenitrificans. Int. J. Syst. Evol. Microbiol.* 51433446

30. M. L Miroshnichenko, H Hippe, E Stackebrandt, N. A Kostrikina, N. A Chernyh, C Jeanthon, T. N Nazina, S. S Belyaev, E. A

Bonch-osmolovskaya, 2001Isolation and characterization of *Thermococcus sibiricus* sp. nov. from a Western Siberia high-temperature oil reservoir. *Extremophiles.*58591

31. G Ravot, M Magot, M. L Fardeau, B. K. C Patel, P Thomas, J. L Garcia, B Ollivier, 1999*Fusibacter paucivorans* gen. nov., sp. nov., an anaerobic, thiosulfate-reducing bacterium from an oil-producing well. *Int. J. Syst. Bacteriol.* 4911411147

32. H Dahle, and N. K Birkeland, 2006*Thermovirga lienii* gen. nov., sp. nov., a novel moderately thermophilic, anaerobic, amino-acid-degrading bacterium isolated from a North Sea oil well. *Int. J. Syst. Evol. Microbiol.* 5615391545

33. R. K Nilsen, and T Torsvik, 1996*Methanococcus thermolithotrophicus* isolated from North sea oil field reservoir water. *Appl. Environ. Microbiol.* 62728731

34. M Magot, 2005Indigenous microbial communities in oil fields. In B. Ollivier and M. Magot, (Eds.) Petroleum microbiology. 2134ASM, Washington, DC.

35. N. L Söhngen, 1913Benzin, Petroleum, Paraffinöl und Paraffin als Kohlenstoff- und Energiequelle für Mikroben. *Zentr Bacteriol Parasitenk Abt II* 37595609

36. F Widdel, and F Musat, 2010Diversity and common principles in enzymatic activation of hydrocarbons. In: K. N. Timmis (Ed.) *Handbook of Hydrocarbon and Lipid Microbiology.* Berlin, Heidelberg: Springer Berlin Heidelberg.

37. M Boll, and J Heider, 2010Anaerobic Degradation of Hydrocarbons: Mechanisms of C-H-Bond activation in the absence of oxygen. In: K. N. Timmis (Ed.) *Handbook of Hydrocarbon and Lipid Microbiology.* Berlin, Heidelberg: Springer Berlin Heidelberg.

38. J Foght, 2008Anaerobic biodegradation of aromatic hydrocarbons: pathways and prospects. *J Mol Microbiol Biotechnol.* 15(2-3): 93-120.

39. J. B Van Beilen, and E. G Funhoff, 2007Alkane hydroxylases involved in microbial alkane degradation. *Appl Microbiol Biotechnol.* 7411321

40. A Wentzel, T. E Ellingsen, H. K Kotlar, S. B Zotchev, M Throne-holst, 2007Bacterial metabolism of long-chain n-alkanes. *Appl Microbiol Biotechnol.* 76612091221

41. F Rojo, 2009*Degradation of alkanes by bacteria. Environmental microbiology.*

42. J. D Van Hamme, A Singh, O. P Ward, 2003Recent advances in petroleum microbiology. *Microbiol Mol Biol Rev.* 674503549

43. F Rojo, 2010Enzymes for Aerobic Degradation of Alkanes. In K. N. Timmis (Ed.), *Handbook of Hydrocarbon and Lipid Microbiology* (781Berlin, Heidelberg: Springer Berlin Heidelberg.

44. R Margesin, D Labbe, F Schinner, C Greer, L Whyte, 2003Characterization of hydrocarbon-degrading microbial populations in contaminated and pristine alpine soils. *Appl Environ Microbiol.* 69630853092

45. E Kuhn, G. S Bellicanta, V. H Pellizari, 2009New alk genes detected in Antarctic marine sediments. *Environ Microbiol.* 113669673

46. J. M Salminen, P. M Tuomi, K. S Jorgensen, 2008Functional gene abundances (*nahAc, alkB, xylE*) in the assessment of the efficacy of bioremediation. *Appl Biochem Biotechnol* 151638652

47. N Hamamura, M Fukui, D. M Ward, W. P Inskeep, 2008Assessing soil microbial populations responding to crude-oil amendment at different temperatures using phylogenetic, functional gene (*alkB*) and physiological analyses. *Environ Sci Technol* 4275807586

48. J. B Van Beilen, M. G Wubbolts, B Witholt, 1994Genetics of alkane oxidation by *Pseudomonas oleovorans*. *Biodegradation* 561174

49. R Marchant, F. H Sharkey, I. M Banat, T. J Rahman, A Perfumo, 2006The degradation of n-hexadecane in soil by thermophilic geobacilli. *FEMS Microbiol Ecol.* 561444

50. J. B Van Beilen, Z Li, W. A Duetz, T. H. M Smits, B Witholt, 2003Diversity of Alkane Hydroxylase Systems in the Environment. *Oil Gas Sci Technol.* 584427440

51. T. P Tourova, T. N Nazina, E. M Mikhailova, T. A Rodionova, A. N Ekimov, A. V Mashukova, A. B Poltaraus, 2008alkB homologs in thermophilic bacteria of the genus *Geobacillus*. *Mol Biol.* 422217226

52. W Li, L. Y Wang, R. Y Duan, J. F Liu, J. D Gu, B. Z Mu, 2012Microbial community characteristics of petroleum reservoir production water amended with n-alkanes and incubated under nitrate-, sulfate-reducing and methanogenic conditions. *Inter Biodeterior Biodegradation.* 698796

53. R Vilchez-vargas, H Junca, D. H Pieper, 2010Metabolic networks, microbial ecology and "omics" technologies: towards understanding in situ biodegradation processes. *Environ Microbiol.* 1230893104

54. L Feng, W Wang, J Cheng, Y Ren, G Zhao, C Gao, Y Tang, et al2007Genome and proteome of long-chain alkane degrading Geobacillus thermodenitrificans NG80-2 isolated from a deep-subsurface oil reservoir. *Proc Natl Acad Sci U S A.* 1041356027

55. M Carmona, M Zamarro, B Blazquez, G Durante-rodriguez, J Juarez, J Valderrama, *et al*2009Anaerobic catabolism of aromatic compounds: a genetic and genomic view. *Microbiol Mol Biol Rev.* 7371133

56. M. V Brennerova, J Josefiova, V Brenner, D. H Pieper, H Junca, 2009Metagenomics reveals diversity and abundance of meta-cleavage pathways in microbial communities from soil highly contaminated with jet fuel under air-sparging bioremediation. *Environ Microbiol.* 119221627

57. D Pérez-pantoja, B González, D. H Pieper, 2010Aerobic degradation of aromatic hydrocarbons. In: K. N. Timmis (Ed.) *Handbook of Hydrocarbon and Lipid Microbiology.* Berlin, Heidelberg: Springer Berlin Heidelberg.

58. Y Jouanneau, 2010Oxidative inactivation of ring cleavage extradiol dioxigenases: mechanism and ferredoxin mediated reactivation. In: K. N. Timmis (Ed.) *Handbook of Hydrocarbon and Lipid Microbiology*. Berlin, Heidelberg: Springer Berlin Heidelberg.

59. H Junca, and D. H Pieper, 2003Functional gene diversity analysis in BTEX contaminated soils by means of PCR-SSCP DNA fingerprinting: comparative diversity assessment against bacterial isolates and PCR-DNA clone libraries. *Environ Microbiol.* 6295110

60. F Mehboob, H Junca, G Schraa, A. J. M Stams, 2009Growth of Pseudomonas chloritidismutans AW-1(T) on n-alkanes with chlorate as electron acceptor. *Appl Microbiol Biotechnol.* 83473947

61. G Fuchs, 2008Anaerobic metabolism of aromatic compounds. *Ann N Y Acad Sci.* 11258299

62. M Kube, J Heider, J Amann, P Hufnagel, S Kühner, A Beck, R Reinhardt, et al2004Genes involved in the anaerobic degradation of toluene in a denitrifying bacterium, strain EbN1. *Arch Microbiol.* 181318294

63. M Boll, G Fuchs, J Heider, 2002Anaerobic oxidation of aromatic compounds and hydrocarbons. *Curr Opin Chem Biol.* 6560411

64. f. M Kaser, and J. D Coates, 2010Nitrate, Perchlorate and Metal respirers. In: K. N. Timmis (Ed.) *Handbook of Hydrocarbon and Lipid Microbiology*. Berlin, Heidelberg: Springer Berlin Heidelberg.

65. O Grundmann, A Behrends, R Rabus, J Amann, T Halder, J Heider, F Widdel, 2008Genes encoding the candidate enzyme for anaerobic activation of n-alkanes in the denitrifying bacterium, strain HxN1. *Environ Microbiol.* 10237685

66. S. R Kane, H. R Beller, T. C Legler, R. T Anderson, 2002Biochemical and genetic evidence of benzylsuccinate synthase in toluene-degrading, ferric iron-reducing *Geobacter metallireducens*. *Biodegradation*, 13214954

67. C Winderl, S Schaefer, T Lueders, 2007Detection of anaerobic toluene and hydrocarbon degraders in contaminated aquifers using benzylsuccinate synthase (bssA) genes as a functional marker. *Environ Microbi*ol 910351046

68. C Winderl, B Anneser, C Griebler, R. U Meckenstock, T Lueders, 2008Depth resolved quantification of anaerobic toluene degraders and aquifer microbial community patterns in distinct redox zones of a tar oil contaminant plume. *Appl Environ Microbiol* 74792801

69. M Staats, M Braster, W. F. M Roling, 2011Molecular diversity and distribution of aromatic hydrocarbon-degrading anaerobes across a landfill leachate plume. *Environ Microbiol* 1312161227

70. B Wawrik, M Mendivelso, V. A Parisi, J. M Suflita, I. A Davidova, C. R Marks, J. D Van Nostrand, Y Liang, J Zhou, B. J Huizinga, et al2012Field and laboratory studies on the bioconversion of coal to methane in the San Juan Basin. *FEMS Microbiol Ecol.* 812642

71. A. V Callaghan, I. A Davidova, K Savage-ashlock, V. A Parisi, L. M Gieg, J. M Suflita, J. J Kukor, et al2010Diversity of benzyl- and alkylsuccinate synthase genes in hydrocarbon-impacted environments and enrichment cultures. *Environ Sci Technol.* 4419728794

72. F Widdel, and O Grundmann, 2010Biochemistry of the anaerobic degradation of non-methane alkanes. In: K. N. Timmis (Ed.) *Handbook of Hydrocarbon and Lipid Microbiology*. Berlin, Heidelberg: Springer Berlin Heidelberg.

73. V Grossi, C Cravolaureau, R Guyoneaud, A Ranchoupeyruse, A Hirschlerrea, 2008Metabolism of n-alkanes and n-alkenes by anacrobic bacteria: A summary. *Org Geochem.* 39811971203

74. L. M Gieg, I. A Davidova, K. E Duncan, J. M Suflita, 2010Methanogenesis, sulfate reduction and crude oil biodegradation in hot Alaskan oilfields.*Environ Microbiol.* 1211307486

75. S. M Mbadinga, K. P Li, L Zhou, L. Y Wang, S Yang, Z Liu, J. F Gu, J.D., et al2012Analysis of alkane-dependent methanogenic community derived from production water of a high-temperature petroleum reservoir. *Appl Microbiol Biotechnol.* 96253142

76. L. M Gieg, and J. M Suflita, 2002Detection of anaerobic metabolites of saturated and aromatic hydrocarbons in petroleum-contaminated aquifers.*Environ. Sci. Technol.* 361737553762

77. K. E Duncan, L. M Gieg, V. A Parisi, R. S Tanner, J. M Suflita, Green Tringe, S., Bristow, J. (2009Biocorrosive thermophilic microbial communities in Alaskan North Slope oil facilities. *Environ Sci Technol* 4379777984

78. A. V Callaghan, B Wawrik, NıChadhain, S.M., Young, L.Y., Zylstra, G.J. (2008Anaerobic alkane-degrading strain AK-01 contains two alkylsuccinate synthase genes. *Biochem Biophys Res Commun.* 366142148

79. J Zedelius, R Rabus, O Grundmann, I Werner, D Brodkorb, F Schreiber, P Ehrenreich, A Behrends, H Wilkes, M Kube, R Reinhardt, F Widdel, 2010Alkane degradation under anoxic conditions by a nitrate-reducing bacterium with possible involvement of the electron acceptor in substrate activation. *Environ Microbiol Rep.* 31125135

80. S. M Mbadinga, L. Y Wang, L Zhou, J. F Liu, J. D Gu, B. Z Mu, 2011Microbial communities involved in anaerobic degradation of alkanes. *Inter Biodeterior Biodegradation.* 651113

81. C So, C Phelps, L Young, 2003Anaerobic transformation of alkanes to fatty acids by a sulfate-reducing bacterium, strain Hxd3. *Appl Environ.* 69738923900

82. A. V Callaghan, M Tierney, C. D Phelps, L. Y Young, 2009Anaerobic biodegradation of n-hexadecane by a nitrate-reducing consortium. *Appl Environ Microbiol* 7513391344

83. Head, I., Gray, N., Aitken, C., Sherry, A., Jones, M., Larter, S. (2010). Hydrocarbon activation under sulfate-reducing and

methanogenic conditions proceeds by different mechanisms. Geophysical Research Abstracts 12 (EGU General Assembly 2010

84. V Torsvik, J Goksoyr, F. L Daae, 1990High diversity in DNA of soil bacteria. *Appl Environ Microbiol* 56782787

85. R. I Amann, W Ludwig, K. H Schleifer, 1995Phylogenetic identification and in situ detection of individual microbial cells without cultivation.*Microbiol Rev* 59143169

86. V Torsvik, F. L Daae, R. A Sandaa, L Øvreås, 1998Novel techniques for analyzing microbial diversity in natural and perturbed environments. *J Biotechnol* 645362

87. E Kellenberger, 2001Exploring the unknown: the silent revolution of microbiology. *EMBO reports*, 2(1), 2-5.

88. V. J Orphan, S. K Goffredi, E. F Delong, J. R Boles, 2003Geochemical influence on diversity and microbial processes in high temperature oil reservoirs. *Geomicrobiol J* 20295311

89. T. N Nazina, N. M Shestakova, Grigor'yan, A.A., Mikhailova, E.M., Tourova, T.P., Poltaraus, A.B., *et al.* (2006Phylogenetic diversity and activity of anaerobic microorganisms of high-temperature horizons of the Dagang oil field (P.R. China). *Microbiology* 755565

90. D Mayumi, H Mochimaru, H Yoshioka, S Sakata, H Maeda, Y Miyagawa, M Ikarashi, et al2011Evidence for syntrophic acetate oxidation coupled to hydrogenotrophic methanogenesis in the high-temperature petroleum reservoir of Yabase oil field (Japan). *Environ Microbiol.* 13819952006

91. J. L Garcia, B. K Patel, B Ollivier, 2000Taxonomic, phylogenetic, and ecological diversity of methanogenic Archaea. *Anaerobe* 6205226

92. S. H Zinder, and M Koch, 1984Non-acetoclastic methanogenesis from acetate: acetate oxidation by a thermophilic syntrophic coculture. *Arch Microbiol* 138263272

93. A Schnurer, F. P Houwen, B. H Svensson, 1994Mesophilic syntrophic acetate oxidation during methane formation by a

triculture at high ammonium concentration. *Arch Microbiol* 1627074

94. S Hattori, Y Kamagata, S Hanada, H Shoun, 2000*Thermacetogenium phaeum* gen. nov., sp. nov., a strictly anaerobic, thermophilic, syntrophic acetate-oxidizing bacterium. *Int J Syst Evol Microbiol 50*16011609

95. M Balk, J Weijma, A. J Stams, 2002*Thermotoga lettingae* sp. nov., a novel thermophilic, methanoldegrading bacterium isolated from a thermophilic anaerobic reactor. *Int J Syst Evol Microbiol 52*13611368

96. T Barth, 1991Organic-acids and inorganic-ions in waters from petroleum reservoirs, Norwegian continental-shelf: a multivariate statistical-analysis and comparison with American reservoir formation waters. *Appl Geochem 6*115

97. T. R Silva, L. C. L Verde, Santos Neto, E.V., Oliveira, V.M. (2012Diversity analyses of microbial communities in petroleum samples from Brazilian oil fields. Inter Biodeterior Biodegradation doi:10.1016/j.ibiod.2012.05.005.

98. J Dolfing, S. R Larter, I. M Head, 2008Thermodynamic constraints on methanogenic crude oil biodegradation. *ISME J 2*442452

99. T Uchiyama, and K Miyazaki, 2009Functional metagenomics for enzyme discovery: challenges to efficient screening. *Curr Opin Biotechnol.20*6616622

100. H Suenaga, T Ohnuki, K Miyazaki, 2007Functional screening of a metagenomic library for genes involved in microbial degradation of aromatic compounds. *Environ Microbiol. 9*922892297

101. H Suenaga, Y Koyama, M Miyakoshi, R Miyazaki, H Yano, M Sota, Y Ohtsubo, et al2009Novel organization of aromatic degradation pathway genes in a microbial community as revealed by metagenomic analysis. *ISME J. 3*12133548

102. I. N Sierra-garcia, Caracterização estrutural e funcional de genes de degradação de hidrocarbonetos originados de

metagenoma microbiano de reservatório de petróleo. M SC. Thesis. Universidade Estadual de Campinas; 2011

103. T Uchiyama, T Abe, T Ikemura, K Watanabe, 2005Substrate-induced gene-expression screening of environmental metagenome libraries for isolation of catabolic genes. *Nat Biotechnol.* 2318893

104. A Ono, R Miyazaki, M Sota, Y Ohtsubo, Y Nagata, M Tsuda, 2007Isolation and characterization of naphthalene-catabolic genes and plasmids from oil-contaminated soil by using two cultivation-independent approaches. *Appl Microbiol Biotechnol.* 74250110

Experimental and Numerical Studies of Reduced Fracture Conductivity Due to Proppant Embedment in the Shale Reservoir

Junjing Zhang[a], Liangchen Ouyang[b], D. Zhu[c], and A.D. Hill[c]

[a]ConocoPhillips [b]PetroChina [c]Texas A&M University

ABSTRACT

Artificially created fracture networks with sufficient fracture conductivities are essential for economic production from shale reservoirs. Fracture conductivity can be significantly reduced in shale formations due to severe proppant embedment. In addition, proppant embedment induces shale flakes that migrate and clog fracture networks.

A laboratory investigation was performed to understand how excessive proppant embedment caused by the shale-water interaction impairs shale fracture conductivity. The experiments were conducted using Barnett shale samples with representative rock properties. The asperities on the fracture surface were carefully preserved. The damage process was simulated in the laboratory by flowing water through the shale fracture packed with proppants. The water used in the experiments had a similar chemical composition to flowback water in the field. The laboratory results were benchmarked with the results from an experimental study conducted with Berea sandstone samples. Post experimental analysis included microscopic imaging of the fracture surfaces and measurement of the proppant embedment depth.

A computational fluid dynamics study was conducted to quantify the conductivity loss due to proppant embedment on a theoretical basis. We developed pore-scale physical models of the proppant pack and calculated the fracture conductivity loss at different proppant embedment depths. The computation was repeated for a variety of proppant layers. The worst case assumed a 40% proppant grain volume embedment.

The experimental study showed up to 88% reduction in fracture conductivity after water flow under 4,000 psi closure stress. The conductivity loss was due to severe proppant embedment as the shale fracture face was softened after its exposure to water. Direct measurement of embedment depths indicated that for fractures that were exposed to water, the average embedment depth was about 50% of the proppant median diameter, while for fractures that were only exposed to gas, the average embedment depth was just 15% of the proppant median diameter. It was also observed that pore space of the sand grains at the outlet of the fracture was clogged by shale flakes and fragments. The computational fluid dynamics study proved that even a 10% proppant grain volume embedment can cause 45%~80% conductivity loss. With the same proppant volume loss due to embedment, the conductivity reduction was less in fractures containing multiple proppant layers than the fracture containing only one layer of proppants.

INTRODUCTION

Extensive studies on realistic fracture conductivity were performed during the past half century. Darin and Huitt (1959) believed that a partial monolayer of proppants can obtain high flow capacity and they theoretically derived an equation to calculate fracture conductivity based on the assumption that spherical proppant grains embed into the formation uniformly. Cooke (1973) studied the effect of high-temperature brine, non-Darcy flow and elevated closure stress on fracture conductivity and determined that these factors can reduce fracture conductivities. The conductivity cell that Cooke used in his study is the prototype of the current API conductivity cell. Cooke (1975) carried on his research to investigate the effect of fracturing fluid. He found that the residue from guar polymer is the most important material in the fracturing fluids that can cause conductivity damage and factors such as proppant concentration, residue of the fracturing fluid, and porosity of the proppant pack determine the conductivity reduction. Following these early works, more studies were done to further understand the damage caused by elevated stress, temperature, fracturing fluid filter cake and fluid additives (Reed, 1980 and Parker and McDaniel, 1987).Fredd et al. (2001) fractured the East Texas Cotton Valley tight sandstone samples and preserved the fracture surface roughness. They measured the conductivities of fractures that were either unpropped or propped by low concentration proppants, and reported that the fracture surface displacement can provide significant unpropped conductivities. Systematic studies on long term fracture conductivity, the impact of fracturing fluid leak-off, gel damage, non-Darcy flow, multiphase flow, proppant crushing and embedment have been done by the Stim-Lab (now a Core-Lab company) Proppant Consortium. Bang et al. (2008) andYuan

(2012) investigated the effective multiphase conductivity loss due to water/condensate block by physical and numerical simulation, and proposed mitigation of liquids block.

The sensitivity of shale to water has been studied in the areas of drilling engineering and formation damage. Common authigenic clay minerals present in petroleum reservoirs are kaolinite, chlorite, illite, smectite and mixed-layer clays (Civan, 2007). The movement of ions/water into and out of shale happens in many ways, including convection, osmosis, capillary imbibition, and diffusion (Yan et al., 2013). Differential pressure between the hydraulic fracture and the matrix can cause water movement. However, convective flow into the matrix is limited due to the ultra-low shale matrix permeability and relatively low differential pressure during a fracturing job. When micro-fractures are induced around the embedded proppant in brittle shale rocks (Kassis and Sondergeld, 2010), water flows through the fractures under differential pressure. Osmosis is considered as an important mechanism of ion migration into and out of the shale matrix because shale acts as a semi-permeable membrane (Low and Anderson, 1958). The osmotic efficiency of clay depends on its porosity, salinity, cation exchange capacity and confining pressure (Fritz and Marine, 1983 and Mody and Hale, 1993). Capillary pressure is the pressure difference across the interface between two immiscible fluids. It is a function of interfacial tension, contact angle and the effective radius. Chenevert and Sharma (1993) believed that the driving force of water movement can be best described by the concept of total aqueous potential differences between shale water and injected water. In recent years, conductivity damage due to clay-water interaction has been brought into attention due to hydraulic fracturing in shale formations. During pre-fracturing formation evaluation in shale, an unpropped fracture conductivity test (UFCT) is sometimes done to determine residual fracture conductivity in the shale rock and different fluid sensitivity. In this test, a horizontal fracture along the shale rock bedding plane is induced in a 1 in. core plug. KCl and NaCl at various concentrations as well as fresh water are injected into the fracture (Ramurthy et al., 2011). This experiment is also run to determine the differential pressure needed to allow the liquid to flow in the fracture. There are limited studies on the conductivity impairment by water for propped fractures under realistic proppant concentrations in shale formations. The water damage mechanisms are yet to be reported with sound laboratory evidence. This work aims at investigating the causes for significant conductivity reduction after water flow in shale reservoirs using both experimental and numerical methods.

EXPERIMENT DESCRIPTION

Shale Sample

The Barnett shale formation was deposited throughout the Fort Worth Basin and tends to thin towards the Llano Uplift in the south with thickness ranging between 30 ft and 50 ft where the outcrop samples were collected (Papazis, 2005 and Zhang et al., 2013). According to X-ray Diffraction tests, the outcrop samples in this study contain 31% quartz, 32% illite, 9% mixed layered illite-smectite and 5% kaolinite (Table 1). Another mineral in the Barnett shale outcrop is anhydrite, which is the main natural fracture infill material (Zhang et al., 2013). The samples were cut into dimensions fit for the modified API conductivity cell. Due to thickness limitation of the outcrop samples, sandstones were cut to make the 1~1.5 in. thick shale samples up to 3 in. which is required by the modified API conductivity cell (Fig. 1).

Table 1: Mineralogy of the Barnett Shale samples by X-Ray Diffraction

Quartz	31%
Feldspar	2%
Kaolinite	5%
Illite	32%
Mixed layered illite-smectite	9%
Chlorite	4%
Other	17%

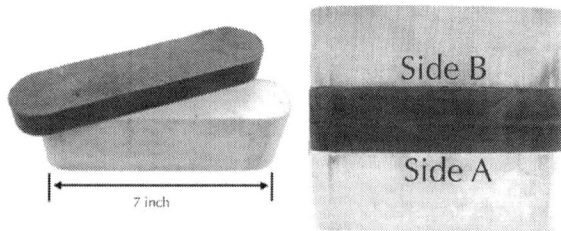

Figure 1: Barnett shale samples shaped to fit into the modified API conductivity cell.

Fluids and Proppants

Industrial grade dry nitrogen was used to measure the undamaged fracture conductivity and the recovered fracture conductivity after water damage. Brine with a similar chemical composition to typical flowback water was injected into the fracture after nitrogen to simulate the damage process. A typical flowback water sample in the Barnett shale has total dissolved solids (TDS) of 39,000 mg/L (Horner et al., 2011). In this study, we formulated the brine with total dissolved solids of 38,000 mg/L (Table 2). 40/70 mesh Badger sand was used in this study. Some quality parameters of the proppants are shown in Table 3. Sieve analysis was done to understand the particle size distribution (Fig. 2).

Table 2: Chemical compositions of the reconstituted water sample in the Barnett shale (Horner et al., 2011)

Ions	Flowback water (mg/L)	Lab (mg/L)
Na^+	12453	12646
Mg^{2+}	253	
Ca^{2+}	2242	2244
Sr^{2+}	357	
Ba^{2+}	42	0.0807

Mn^{2+}	44	0.0066
Fe^{2+}	33	
SO_4^{2-}	60	
HCO_3^-	289	9
Cl^-	23797.5	23170
TDS	39,570	38,069

Table 3: Quality parameters of the proppants

Grain diameter (mm)	Nephelometric Turbidity Units (NTU)	Roundness	Sphericity
0.149~0.595	53	0.75	0.75

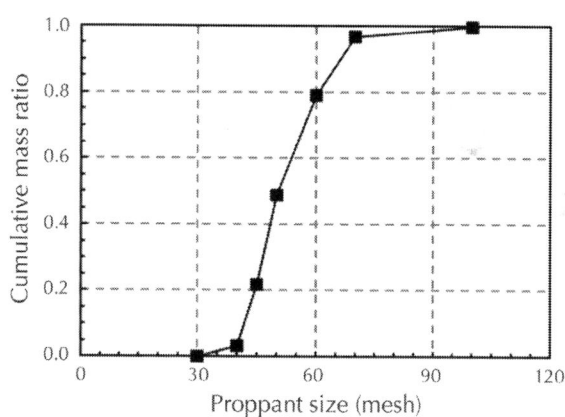

Figure 2: Particle size distribution of 40/70 mesh sands.

Experimental Setup

The entire apparatus consists of five separate units: (1) gas injection unit; (2) liquid injection unit; (3) conductivity cell assembly; (4) closure stress application unit; and (5) pressure/rate data acquisition unit.

The schematic of the setup is shown in Fig. 3. The gas injection unit consists of a nitrogen tank containing industrial grade dry nitrogen and a pressure regulator. The liquid injection unit includes a 1000D Teledyne ISCO Syringe pump, a PVC refilling accumulator, two stainless steel displacement accumulators, an AW-32 hydraulic oil reservoir and the flow line manifolds. The 1000D Teledyne syringe pump has one liter nominal capacity. The flow rate range for this pump is 0.1~408 mL/min and the working pressure is 0~2,000 psi. In Fig. 3, the parts highlighted in orange are only in contact with AW-32 hydraulic oil. All other aspects of the modified conductivity cell are designed as per the API conductivity cell except the dimensions. The modified API conductivity cell accommodates 6 in. thick rock samples. The hydraulic load frame can apply 208,000 lbf at the rate of 1,215 lbf/min (stress rate 100 psi/min). The position sensor of the load frame has an accuracy of 4×10^{-4} in. The gas flow controller has a full range of 0.35 ft^3/min with an accuracy of 3.5×10^{-5} ft^3/min. Differential pressure sensor diaphragms can be switched and the pressure ratings used in this study are 5 psi and 20 psi.

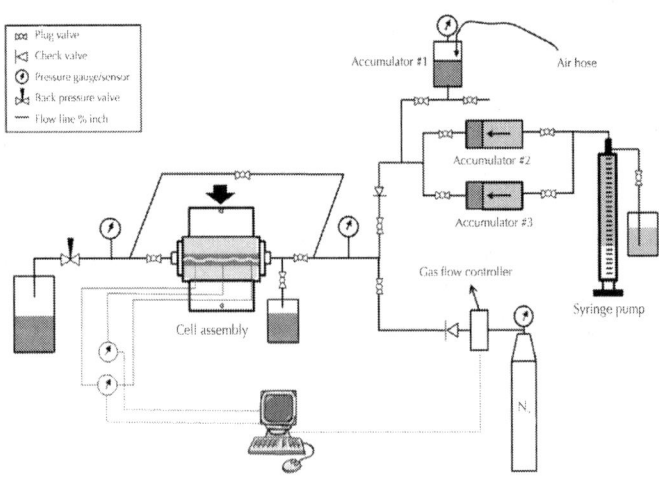

Figure 3: Schematic of the setup to evaluate the conductivity damage by water.

Experimental Procedures and Conditions

To evaluate the conductivity damage, both gas and water were flowed through the fracture. Dry nitrogen was flowed first to measure the undamaged fracture conductivity since nitrogen does not react with shale. Then, water was injected into the fracture to soak the shale fracture faces until the steady state flow was reached. Finally, dry nitrogen was flowed again to remove water from the fracture and to measure the recovered fracture conductivity. Severity of conductivity damage by water was assessed by comparing the two gas-measured conductivities.

Experiments were run at 70 °F. Both proppants and shale samples were kept dry before experiment. The undamaged fracture conductivity was measured in the short-term experiment as per API RP-61. Water was injected at the higher rate (0.5 mL/min) to saturate the system and then at the lower rate (0.1 mL/min) until the flow stabilized. The second gas flow removed water from the fracture by displacement and evaporation. Gas flowed through the fracture until the steady state regime appeared. The effective closure stress acting on the fracture was set at 4,000 psi. Calculation of the effective closure stress can be found inZhang et al., (2013).

Post-Experimental Analysis

Images of the fracture faces were taken after the experiments using a Zeiss Axiophot microscope. An experiment where the shale samples were only exposed to dry gas was also run. Direct comparison of the fracture faces between the samples exposed to water and only to dry gas was made first; then a quantitative evaluation of the proppant embedment with and without exposure to water was performed.

NUMERICAL METHODOLOGY

A series of numerical simulation cases to study the effect of proppant embedment on the conductivity reduction were performed. The conductivity reduction due to proppant embedment was modeled with a computational fluid dynamics approach. Recently, pore-scale models in CFD have been widely used in studying fluid flow through porous media (Jiang and Lu, 2005, Xu and Jiang, 2008 and Ouyang et al., 2013). We used the CFD software package FLUENT (ANSYS Inc.) in this work. It can account for complicated pore scale geometries, and is cost effective for large computations.

Governing Equations

FLUENT uses the finite-volume method to solve the three-dimensional Navier-Stokes equations. Consistent with the experimental conditions for conductivity measurements, the flow was steady state and under 70 °F. The continuity and momentum balance equations for the steady state flow are shown below.

Continuity equation

$$\frac{\partial u}{\partial x} + \frac{\partial v}{\partial y} + \frac{\partial w}{\partial z} = 0$$

(1)

Momentum equations

$$\rho\left(u\frac{\partial u}{\partial x} + v\frac{\partial u}{\partial y} + w\frac{\partial u}{\partial z}\right) = \frac{\partial P}{\partial x} + \mu\left(\frac{\partial^2 u}{\partial x^2} + \frac{\partial^2 u}{\partial y^2} + \frac{\partial^2 u}{\partial z^2}\right)$$

$$\rho\left(u\frac{\partial v}{\partial x} + v\frac{\partial v}{\partial y} + w\frac{\partial v}{\partial z}\right) = \frac{\partial P}{\partial y} + \mu\left(\frac{\partial^2 v}{\partial x^2} + \frac{\partial^2 v}{\partial y^2} + \frac{\partial^2 v}{\partial z^2}\right)$$

$$\rho\left(u\frac{\partial w}{\partial x} + v\frac{\partial w}{\partial y} + w\frac{\partial w}{\partial z}\right) = \frac{\partial P}{\partial z} + \mu\left(\frac{\partial^2 w}{\partial x^2} + \frac{\partial^2 w}{\partial y^2} + \frac{\partial^2 w}{\partial z^2}\right)$$

(2)

Computational Domain

Computationally, numerical simulation of fluid flow inside the field scale fracture geometry using CFD exceeds the capacity of most computers. To reduce the computational load but still keep the simulations representative, the computational domain shown in Fig. 4(a) was used. Besides the three layers of proppants, the domain also contains an entrance and an exit region of four proppant diameters length on each side. The extended regions are to avoid the impact of inflow and outflow on calculated results and to reduce the possibility of counter-flow during the computation. The proppant particles are arranged in body centered cubic pattern as shown in Fig. 4(b). They form a homogeneous and uniform porous media packing. Four different proppant sizes (20 mesh, 40 mesh, 70 mesh and 100 mesh) are used in order to study the effect of proppant size on fracture conductivity. To mimic the realistic situation, the surfaces of adjacent proppants have finite contact points. Gas flows from left to right. The top and bottom surfaces represent the shale fracture boundaries. The domain is created by GAMBIT (ANSYS Inc.), a general-purpose preprocessing CFD software used to generate flow geometry and meshes.

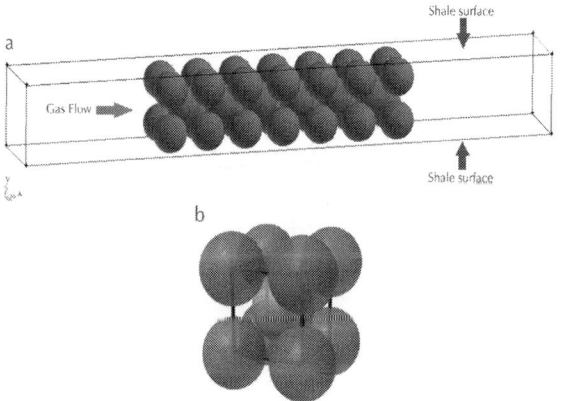

Figure 4: (a) Schematic of the computational domain (b) Geometrical pattern for body centered cubic packing.

Proppant embedment into the shale fracture surface happens with increasing effective closure stress, and is worsened due to fracture surface softening by water. This phenomenon has been represented in the model, as shown in Fig. 5. In Fig. 5(a) and (c), proppants of the top and bottom layers are truncated because embedment into the shale fracture surfaces has occurred. To investigate how the loss of flow area influences the overall fracture conductivity, we modeled multiple cases for one, two and three layers of proppants. 40% of the proppant grain volume is lost due to embedment in the case where the maximum simulated embedment happens.

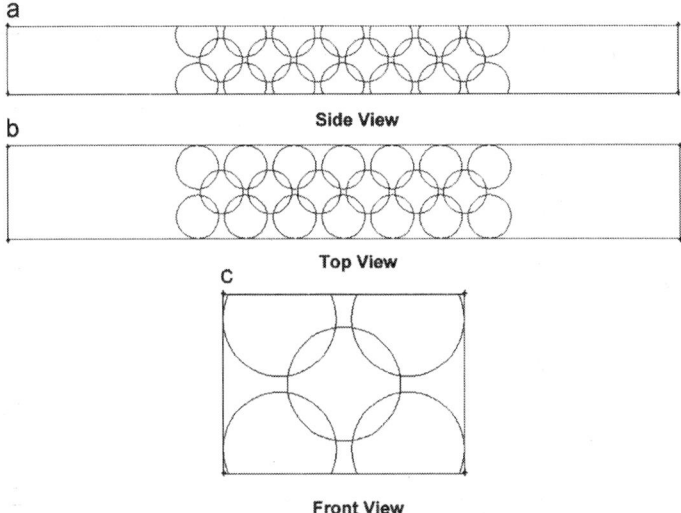

Figure 5: Schematics of the proppant embedment into shale fracture surfaces. (a) Side View (b) Top View (c) Front View.

Numerical Setup

Grid size affects the accuracy of the numerical simulation. We conducted a series of calculations with different grid sizes to check the computational stability and accuracy. To keep a balance between accuracy and efficiency, the grid size of 0.02 mm in the x, y, and z-directions was chosen in all computational cases. Additionally, the numerical simulation cases using a grid size of 0.02 mm did not have any instability problems.

The numerical simulations assumed that the flow was three-dimensional, single-phase and laminar. The second-order upwind scheme was adopted for the spatial discretization of flow pressure and momentum governing equations. The gradient of spatial discretization was based on Green-Gauss Cell. To couple the pressure and the velocity, the Semi-Implicit Method for Pressure Linked Equations (SIMPLE) algorithm was used (Patankar, 1980). A constant velocity boundary condition was imposed at the inlet. A constant zero static pressure condition was used at the outlet, which was relative to the reference cell pressure of 50 psi. Wall boundary conditions were set on the shale surfaces as well as the proppant surfaces as shown in Fig. 4(a). The symmetrical boundary condition was used in the Z direction in Fig. 5(c) to keep periodicity in the cell width direction. The solution was assumed to have converged when the root mean square of the normalized residual error reached 10^{-9}. The cases were run on an IBM iDataplex Cluster with nodes based on Intel's 64-bit Nehalem & Westmere processor.

RESULTS AND DISCUSSIONS

Fracture conductivities of Barnett shale samples were measured by gas before and after the water injection. To benchmark the measurements, fracture conductivities of Berea sandstone samples were also measured under the same experimental conditions. Finally, fracture conductivity reductions due to proppant embedment were calculated using the numerical model to verify the experimental results.

Conductivities in Barnett Shale

The fracture was propped by 40/70 mesh sands at an areal concentration of 0.10 lb/ft². The closure stress was constant at 4,000 psi throughout the experiment. Results are shown in Fig. 6. After the initial measurement by gas, fracture conductivity kept dropping as water was injected displacing gas out of the fracture. After 2 h of water injection at the rate of 0.1 mL/min, the flow reached steady state where the fracture conductivity was calculated. The process of water removal by gas includes two stages: (1) water displacement by gas and (2) evaporation of water. The latter stage is also referred to as "flow-through drying" in porous media (Mahadevan and Sharma, 2003; Zhang et al. 2013b). The two-stage water removal is reflected in the recovered conductivity curve after 370 min. The fast conductivity recovery before 380 min represents the displacement process. During this phase, we observed a continuous water stream for a short period followed by large water droplets coming from the outlet of the flowline. After that, it enters the long evaporation process where the conductivity is recovered slowly until the steady flow condition is reached. During this phase, no large water droplets were seen. The long term gas flow dries the fracture surface so that single phase gas flows through the fracture.

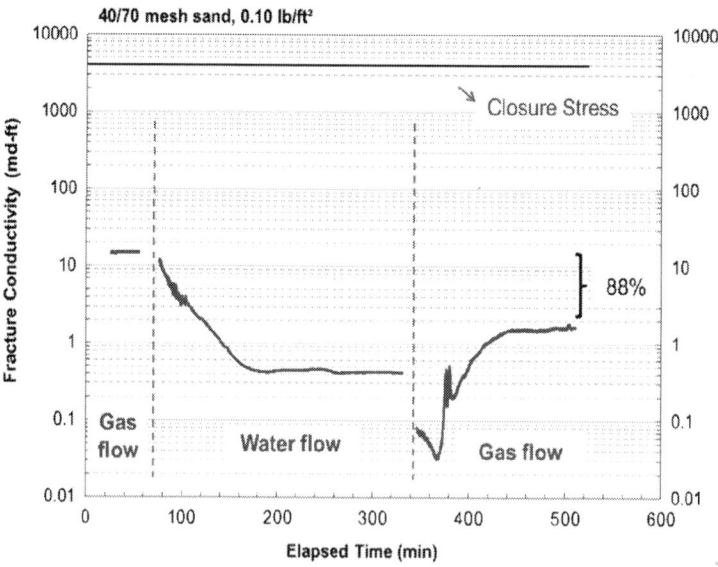

Figure 6: Fracture conductivities of the Barnett shale measured before and after water flow.

The initial undamaged conductivity for the shale fracture is 15 md-ft, which is significantly reduced to 0.11 md-ft after the water flow. The gas flow can only recover the fracture conductivity to 1.7 md-ft. Due to the water flow, there is an 88% unrecoverable conductivity loss at the end of the experiment. Longer term rock creep and proppant rearrangement play an important role in fracture conductivity measurement over time.Zhang et al. (2014) found out that in the first 20 h fracture conductivity can be reduced by 20% under the same experimental conditions while the conductivity variation is negligible in the following 30 h.

More experiments were conducted under the same conditions and similar trends were observed.

Conductivities in Berea Sandstone

Quartz comprises up to 87% of the Berea sandstone and clay only accounts for 6% as shown in Table 4. The same procedures and experimental conditions were applied in the conductivity measurements of sandstone samples. Results are shown in Fig. 7. In this plot, conductivities of Barnett shale samples are from Fig. 6. The initial undamaged fracture conductivity using Berea sandstone samples is 65 md-ft. After water flow, the recovered fracture conductivity by gas is 61 md-ft. The 94% conductivity recovery after water flow indicates that there is negligible damage to conductivity by water in the Berea sandstone fracture.

Table 4: Mineralogy of the Berea sandstone samples by X-Ray Diffraction

Quartz	87%
Clay	6%
Carbonate	2%
Other	5%

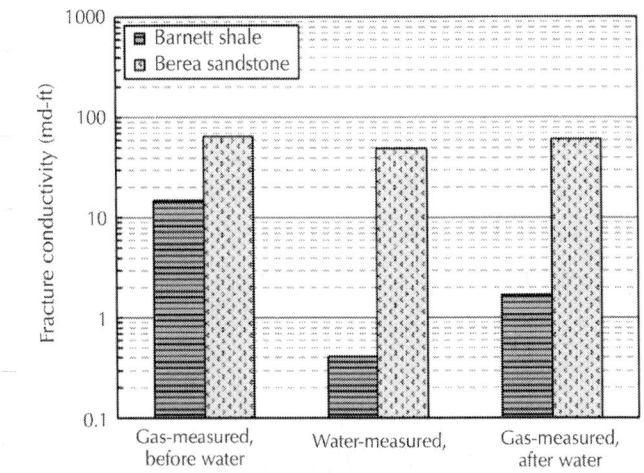

Figure 7: Fracture conductivities of Berea sandstone samples and Barnett shale samples measured before and after water flow. Results of Barnett shale samples are from Fig. 6.

Comparison of the fracture conductivities of Barnett shale samples with Berea sandstone samples clearly suggests that the rock lithology has a significant impact on the recovered fracture conductivity. The clay-water interaction results in severe damage to fracture conductivity in shale.

Shale Fracture Surface Softening

Experimental results show that 88% of the undamaged fracture conductivity is permanently lost in Barnett shale samples. Significant conductivity loss can be attributed to the excessive proppant embedment due to the shale fracture surface softening.

Microscopic images of proppant embedment were taken for both experiments with and without water flow.Fig. 8(a) shows the "moon-surface-like" image of a shale fracture face under oblique light after the water experiment. The embedment craters congregate on the fracture surface next to each other. In Fig. 8(b), some proppants are completely buried into the fracture face. Fig. 9

shows the Barnett shale fracture that is only exposed to gas during the conductivity measurement. Obviously, the embedment is much less and shallower than the fracture exposed to water.

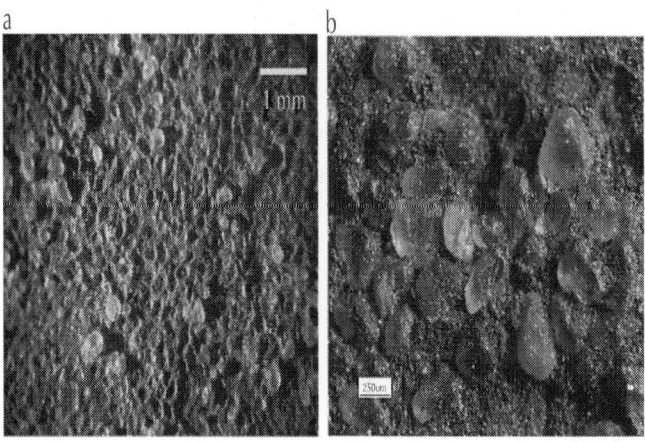

Figure 8: Microscopic images of 40/70 mesh sand embedment into the Barnett shale fracture surface after the conductivity measurement by water. The closure stress was 4,000 psi.

Figure 9: Microscopic image of 40/70 mesh sand embedment into the Barnett shale fracture surface after the conductivity measurement by gas only. The closure stress was 4,000 psi.

Proppant embedment depth was then measured by a microscope at five locations spaced by one inch (Fig. 10). At each location, multiple embedment depths were measured. This image also highlights the fracture surface roughness. The red and blue regions represent the surface hills and valleys respectively. During the measurement, the focus of the microscope was adjusted from the bottom of the embedment crater to the surrounding by an increment of 10 μm. The embedment depth was also measured in the experiment where only gas was flowed. The comparison is shown in Fig. 11.

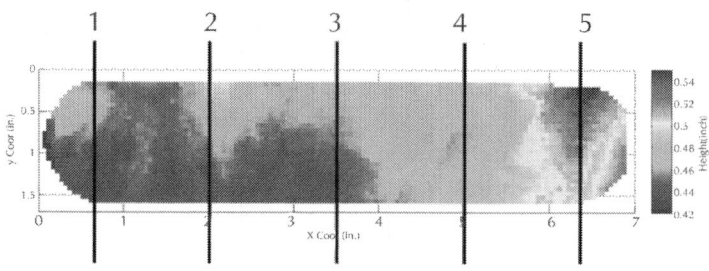

Figure 10: Measurement locations on the shale fracture surface.

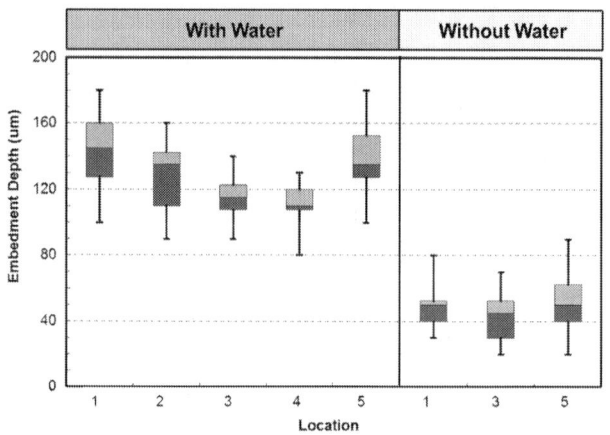

Figure 11: Comparison of the embedment depth after a water test with the embedment depth after a dry gas test. 40/70 mesh sands were placed

on both fractures. The boxplot shows the maximum, 75%, 50%, 25% and the minimum embedment depths. The closure stress was 4,000 psi.

The average embedment depth for the fracture surface after water flow is about 140 μm, while the average embedment depth for the fracture only exposed to gas is just 40 μm. The median diameter of 40/70 mesh sand is 300 μm. This means, on average, half of the sand grain is buried into the fracture face after it is exposed to water due to the shale surface softening.

Shale softening by water has been studied to overcome the challenges in drilling through shale formations. Shale-water interaction in the reservoir leads to reduced effective stress, Young›s modulus, uniaxial compressive strength and eventually causes rock failure (Chenevert and Sharma, 1993, Chen and Ewy, 2002 and Lin and Lai, 2013). When water migrates into clay structure, the local pore pressure is increased. The excessive pore pressure is hard to dissipate due to the ultra-low permeability of the shale matrix (Zhang, 2005). Therefore, the effective stress of the shale matrix where the water front has reached is reduced. This localized pore pressure increment due to water movement is called "undrained condition" (Detournay and Cheng, 1988). Migrated water in the clay lattice causes clay expansion and reduces the interlayer bonding strength of clay (Zhang, 2005). The combination of elevated local pore pressure and reduced strength leads to the softening of shale after being soaked in water.

Fracture Conductivities Calculated by the Numerical Model

In the numerical study, a large range of fracture conditions were investigated. The proppant diameter was varied from 149 μm to 840 μm (100 mesh to 20 mesh) and the fraction of the proppant grain embedment volume was varied from 0 to 0.4. Conductivities of fractures propped by one, two and three layers of proppants were calculated respectively. The results are shown in Fig. 12. Obviously, fracture conductivities are affected by the proppant size, the embedment volume ratio and the number of proppant layers. As shown in Fig. 12(a), for one layer of proppants without embedment, fracture conductivity of 20 mesh proppants is 1406 md-ft while the conductivity of 100 mesh proppants is only 7.8 md-ft under the same conditions. For one layer of 20 mesh proppants, the conductivity drops from 1406 md-ft to 8.3 md-ft when the grain embedment volume ratio increases from 0 to 0.4. The conductivities of two layers and three layers of proppants are significantly improved comparing with just one layer of proppants under the same flow conditions as shown in Fig. 12(b) and (c).

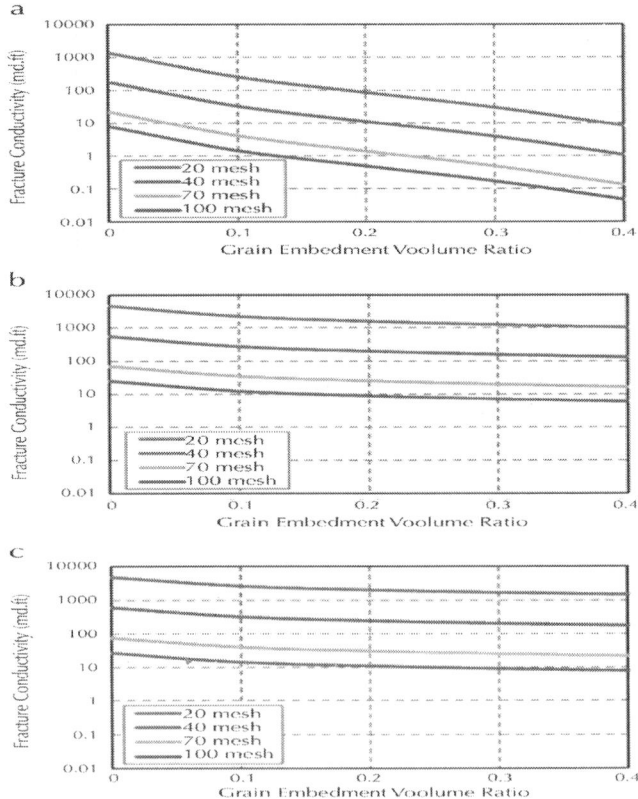

Figure 12: Fracture conductivities as a function of grain embedment volume ratio, proppant size and the number of proppant layers. (a) One layer of proppants (b) Two layers of proppants (c) Three layers of proppants.

Effect of Proppant Embedment and Number of Layers

Fig. 13 shows the grain embedment volume ratio versus the ratio of remaining fracture conductivity for various proppant layers. It can be seen that for a monolayer proppant pack, the fracture conductivity decreases dramatically as the embedment volume ratio increases. As the volume ratio reaches 0.4, 99.4% of the conductivity is lost due to the flow area loss by proppant embedment. This illustrates that proppant embedment can be a major cause for significant shale fracture conductivity loss when low concentration proppant stages are pumped, because monolayer proppant packs are more likely to be created. Two or more layers of proppants can greatly impede the conductivity reduction as a result of proppant embedment. This is because in multiple proppant layers, the conductivity depends more on the internal cleaner proppant pack and less on the flow areas adjacent to the fracture walls.

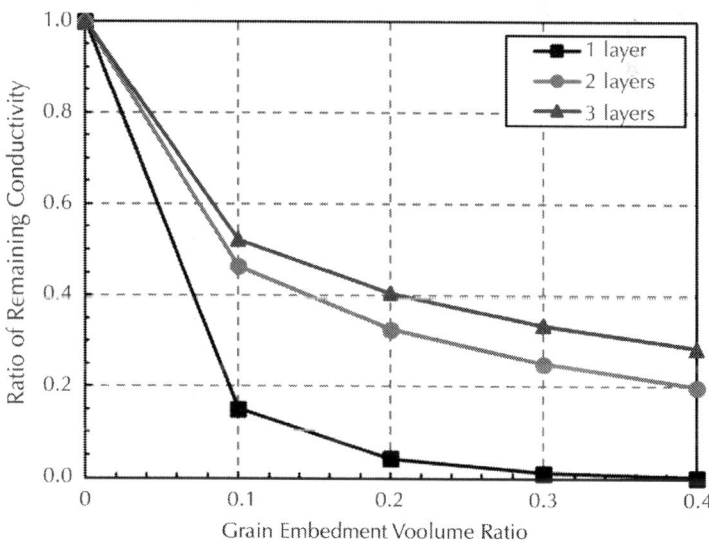

Figure 13: Grain embedment volume ratio versus ratio of remaining fracture conductivities for various proppant layers.

The huge reduction in fracture conductivity at the first 10% proppant grain volume loss can be illustrated byFig. 14. The location near the shale fracture has more open space for fluid flow than the location in the middle, as highlighted in red in Fig. 14(b). For a given grain volume loss due to embedment, there is greater embedment depth in the early embedment phase than the later phase because of the spherical shape of the proppant.

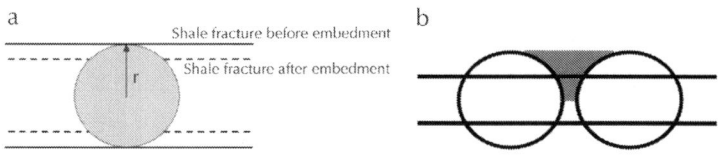

Figure 14: Flow area of the proppant layer close to the fracture walls.

CONCLUSIONS

We have conducted a laboratory study on the mechanism of shale fracture conductivity damage caused by shale-water interactions. Pore scale numerical models were built to enhance the understanding of the effect of proppant embedment on shale fracture conductivity reduction. Based on the experimental and numerical studies, the following conclusions are made.

- Up to 88% of the undamaged shale fracture conductivity was lost after water flow in the Barnett shale under 4,000 psi closure stress. The significant conductivity loss was caused by the shale fracture surface softening after its exposure to water.
- The average embedment depth was about 50% of the proppant median diameter in fractures that were exposed to water, while the average embedment depth was just 15% of the proppant median diameter in fractures that were only exposed to gas.
- The computational fluid dynamics study showed that the first 10% of proppant grain embedment volume can cause

a significant conductivity reduction (45%~80%) depending on the number of proppant layers. Given the same proppant volume loss due to embedment, the conductivity reduction was less in fractures containing multiple proppant layers because with multiple layers of proppants, the fracture conductivity depends more on the internal, cleaner proppants and less on the flow areas adjacent to the fracture walls.

- The numerical study indicated that fractures propped by a monolayer of proppants experienced severe conductivity reduction due to the large loss of flow area even at a small grain embedment volume.

- The pore scale numerical model was able to capture the flow characteristics inside the proppant pack. It proved to be a useful tool for studying the flow behavior including conductivity creation and impairment in hydraulic fractures.

ACKNOWLEDGEMENTS

The authors would like to acknowledge the sponsors of Crisman Institute for Petroleum Research in the Petroleum Engineering Department at Texas A&M University as well as the Department of Energy, and the Research Partnership to Secure Energy for America (RPSEA) for their support of this work (RPSEA project 11122-07-TAMU-Zhu).

REFERENCES

1. ANSYS FLUENT User's Guide, 2010ANSYS FLUENT User's Guide, Release 13.0. ANSYS Inc., Canonsburg, PA 2010.

2. ANSYS GAMBIT User's Guide, 2006 ANSYS GAMBIT User's Guide, Release 2.3. ANSYS Inc., Canonsburg, PA 2006.

3. Bang et al., 2008 Bang, V., Yuan, C., Pope, G.A., Sharma, M.M., Baran, R. Jr., Skildum, J. and Linnemeyer, H.C. 2008. Improving Productivity of Hydraulically Fractured Gas

Condensate Wells by Chemical Treatment. Paper OTC 19599 presented at the SPE Offshore Technology Conference held in Houston, Texas, U.S. A., 5-8 May.

4. Chen and Ewy, 2002 Chen, G. and Ewy, R.T. 2002. Investigation of the Undrained Loading Effect and Chemical Effect on Shale Stability. Paper SPE 78164 presented at the SPE/ISRM Rock Mechanics Conference, Irving, Texas, U.S.A., 20-23 October.

5. Chenevert and Sharma, 1993 M.E. Chenevert, A.K. Sharma Permeability and Effective Pore Pressure of Shales SPE Drilling & Completion, 8 (1) (1993), pp. 28–34

6. Civan, 2007 F. Civan Reservoir Formation Damage Gulf Publishing Company, Houston, Texas (2007)

7. Cooke, 1973 C.E. Cooke Jr.Conductivity of Fracture Proppants in Multiple Layers Journal of Petroleum Technology, 25 (9) (1973), pp. 1101–1107

8. Cooke, 1975 C.E. Cooke Jr. Effect of Fracturing Fluids on Fracture Conductivity Journal of Petroleum Technology, 27 (10) (1975), pp. 1273–1282

9. Darin and Huitt, 1959 Darin, S.R. and Huitt, J.L. 1959. Effect of a Partial Monolayer of Proppant Agent on Fracture Flow Capacity. Paper SPE 1291 presented at the Annual Fall Meeting, Dallas, Texas, U.S.A., 4-7 October.

10. Detournay and Cheng, 1988 E. Detournay, A.H-D. Cheng Poroelastic Response of a Borehole in a Non-Hydrostatic Stress field International Journal of Rock Mechanics and Mining Sciences, 25 (3) (1988), pp. 171–182

11. Fredd et al., 2001 C.N. Fredd, S.B. McConnell, C.L. Boney, K.W. England Experimental Study of Fracture Conductivity for Water-Fracturing and Conventional Fracturing Applications SPE Journal, 6 (3) (2001), pp. 288–298

12. Fritz and Marine, 1983 S.J. Fritz, I.W. Marine Experimental Support for a Predictive Osmotic Model of Clay Membranes Geochemistry Cosmochim, ACTA, 47 (8) (1983), pp. 1515–1522

13. Horner et al., 2011Horner, P., Halldorson, B., and Slutz, J.A. 2011. Shale Gas Water Treatment Value Chain-A Review of Technologies, including Case Studies. Paper SPE 147264 presented at the SPE Annual Technical Conference and Exhibition, Denver, Colorada, U.S.A., 30 October-2 November.

14. Jiang and Lu, 2005 P. Jiang, X. Lu Numerical Simulation of Fluid Flow and Convection Heat Transfer in Sintered Porous Plate Channels International Journal of Heat and Mass Transfer, Volume 49 (Number 9-10) (2005), pp. 1685–1695

15. Kassis and Sondergeld, 2010 Kassis, S. and Sondergeld, C. 2010. Fracture Permeability of Gas Shale: Effects of Roughness, Fracture offset, Proppant, and Effective Stress. Paper SPE 131376 presented at the CPS/SPE International Oil & Gas Conference and Exhibition, Beijing, China, 8-10 June.

16. Lin and Lai, 2013 Lin, S. and Lai, B. 2013. Experimental Investigation of Water Saturation Effects on Barnett Shale's Geomechanical Behaviors. Paper SPE 166234 presented at the SPE Annual Technical Conference and Exhibition, New Orleans, Louisiana, U.S.A., 30 September-2 October.

17. Low and Anderson, 1958 P.F. Low, D.M. Anderson Osmotic Pressure Equation for Determining Thermodynamic Properties of Soil Water Soil Science, 86 (5) (1958), pp. 251–253

18. Mahadevan and Sharma, 2003 Mahadevan, J. and Sharma, M.M. 2003. Clean-up of Water Blocks in Low Permeability Formations. Paper SPE 84216 presented at the SPE Annual Technical Conference and Exhibition, Denver, Colorado, U.S.A., 5-8 October.

19. Mody and Hale, 1993 Mody, F.K. and Hale, A.H. 1993. A Borehole Stability Model to Couple the Mechanics and Chemistry of Drilling Fluid Shale Interaction. Paper SPE/IADC 25728 presented at the 1993 SPE/IADC Drilling Conference, Amsterdam, the Netherlands, 23-25 February.

20. Ouyang et al., 2013 Ouyang, L., Zhu, D. and Hill, A.D. 2013. Effect of Pore Structure on Non-Newtonian Fluid Flow.

Paper IPTC 17011 presented at the International Petroleum Technology Conference, Beijing, China, 26-28 March.

21.   Papazis, 2005 P.K. Papazis Petrographic Characterization of the Barnett Shale, Fort Worth MS Thesis The University of Texas at Austin, Austin, Texas, Basin, Texas (2005) August 2005

22.   Parker and McDaniel, 1987 Parker, M.A. and McDaniel, B.W. 1987. Fracturing Treatment Design Improved by Conductivity Measurements Under In-Situ Conditions. Paper SPE 16907 presented at the SPE Annual Technical Conference and Exhibition, Dallas, Texas, U.S.A., 27-30 September.

23.   Patankar, 1980 S.V. Patankar Numerical Heat Transfer and Fluid Flow Hemisphere Publishing Corporation, Washington, D.C. (1980)

24.   Ramurthy et al., 2011 M. Ramurthy, R.D. Barree, D.P. Kundert, E. Petre, M. Mullen Surface-Area vs Conductivity-Type Fracture Treatments in Shale Reservoirs SPE Production and Operations, 26 (4) (2011), pp. 357–367

25.   Reed, 1980 M.G. Reed Gravel Pack and Formation Sandstone Dissolution during Steam Injection Journal of Petroleum Technology, 32 (6) (1980), pp. 941–949

26.   Xu and Jiang, 2008 R. Xu, P. Jiang Numerical Simulation of Fluid Flow in Microporous Media International Journal of Heat and Mass Transfer, Volume 29 (Number 5) (2008), pp. 1447–1455

27.   Yan et al., 2013 Yan, B., Wang, Y., and Killough, J.E. 2013. Beyond Dual-Porosity Modeling for the Simulation of Complex Flow Mechanisms in Shale Reservoirs. Paper SPE 163651 presented at the SPE Reservoir Simulation Symposium in the Woodlands, Texas, U.S.A., 18-20 February.

28.   Yuan, 2012 Yuan, C. 2012. The Chemical Treatment to Remove Liquid Block in Hydraulic Fractured Well-A Simulation Study with Leak-Off. Paper SPE 158190 presented at the 2012 Energy Conference-Developing Resources for Sustainability at Port-of-Spain, Trinidad, 11-13 June.

29. Zhang, 2005 J. Zhang The Impact of Shale Properties on Wellbore Stability PhD dissertation The University of Texas at Austin, Austin, Texas (2005) August 2005

30. Zhang et al., 2014 J. Zhang, A. Kamenov, D. Zhu, A.D. Hill Laboratory Measurement of Hydraulic Fracture Conductivities in the Barnett Shale SPE Production & Operations, 29 (3) (2014), pp. 216–227

31. Zhang et al., 2013 Zhang, Q., Zhu, D., and Hill, A.D. 2013. Modeling of Spent-Acid Blockage Damage in Stimulated Gas Wells. Paper IPTC 16481 presented at the International Petroleum Technology Conference in Beijing, China, 26-28 March.

Chapter 5

Robust Intelligent Tool for Estimating Dew Point Pressure in Retrograded Condensate Gas Reservoirs: Application of Particle Swarm Optimization

Mohammad Ali Ahmadi[a], Mohammad Ebadi[b], and Arash Yazdanpanah[c]

[a]Department of Petroleum Engineering, Ahwaz Faculty of Petroleum Engineering, Petroleum University of Technology, Ahwaz, Iran

[b]Department of Petroleum Engineering, Science and Research Branch, Islamic Azad University, Tehran, Iran

[c]Iranian Offshore Oil Company (IOOC), Tehran, Iran

ABSTRACT

Liquid production from gas condensate reservoirs, which is an important economic and technical issue, depends on the thermodynamic conditions underlying the porous media. Accurately estimating the relevant parameters is an incentive for researchers to develop and propose a diversity of correlations; however, certain correlations are not sufficiently precise compared with correlations that are routinely applied to determine the dew point pressure (P_d). Due to numerous misunderstandings in P_d estimations, which are typically observed in upstream industries, great effort was expended herein to produce a high-performance method to monitor the P_d. The solution was produced by creating a hybrid of two effective and robust methods, the swarm intelligence and artificial neural network (ANN) models. The proposed model was extended using precise dew point pressure data reported in previous studies; moreover, based on these data, the evolved intelligent approach and conventional schemes were compared. The statistical results show a notable performance by the smart model in determining the dew point pressure of condensate gas reservoirs. Based on the reliable results, which are highly accurate and effective, it can logically be inferred that implementing the proposed approach, PSO-ANN, can aid in better understanding reservoir fluid behavior through reservoir simulation scenarios.

INTRODUCTION

Gas condensate reservoirs, which are one of the most valuable types of hydrocarbon sources with an ability to supplement high levels of energy, have become popular (Mohammadi et al., 2013, Mokhtari et al., 2013 and Sadeghi Booghar and Masihi, 2010). Consequently, preparing effective, complete and multidisciplinary approaches to production using such reservoirs has practical, methodical and monetary significance. To develop vital and crucial plan for production using the aforementioned resources, demands for an exact, accurate and definite understanding of reservoir fluid properties have continuously been considered. In other words, PVT properties, for which even insignificant and infinitesimal mathematical errors in their estimation can result in severe problems for successive processes, are the most important pieces of information that can be determined from each step in reservoir simulations and development (Alavi et al., 2010, Mokhtari et al., 2013, Ursin, 2004 and Yong et al., 2010).

A step referred to as "flow-in" is the starting point for reservoir pressure reduction, which leads to liquid drop formation in zones near the wellbore due to overtaking a pressure threshold referred to as *dew point pressure* (P_d) (Brown et al., 2009, Elsharkawy and Foda, 1998 and Nasrifar et al., 2005). Dramatic reductions in the gas relative permeability and gas production rate are intense effects from the aforementioned drop formation (App and Burger, 2009, Chowdhury et al., 2008 and Thomas et al., 2009). Accordingly, careful determination of the P_d is essential. Hence, great effort either via theoretical or laboratorial methods has been expended, and suggestions have been generated for measuring the P_d (Berning, 2012, Louli et al., 2012, Rahimpour et al., 2011 and Nowroozi et al., 2009). Constant volume depletion (CVD) is an experimental process, which generates Pd from samples. These methodologies, the comprehensive steps for which have been fully determined in previous studies, routinely include technical hitches, such as price and slow speed, and their reliability is effected by external influences, such as human error (Luo et al., 2001, Shadizadeh et al., 2006, Shen et al., 2001, Zheng et al., 2000 and Jalili et al., 2007).

Furthermore, certain mathematically inspired solutions and concepts, such as equation of states (EOS), and empirically derived correlations have been presented that measure the critical PVT properties (Bonyadi and Esmaeilzadeh, 2007, Elsharkawy, 2002a, Elsharkawy, 2002b and Li et al., 2012). For instance, a formula mainly based on C7+ characterization, temperature and fluid composition was generated by Nemeth and Kennedy (1966) through multiple regressions and was supported by an extensive database to predict the P_d. The applicability of the formula is reliable under specific thermodynamic ranges (Nemeth and Kennedy, 1967). Additionally, researchers have attempted to model production from gas condensate reservoirs without using PVT data; this was exemplified in a study by Marruffo et al., who generated an approach for estimating the P_d and C7+ content in gas condensate reservoirs (Marruffo et al., 2001). In addition, the impact of impurities (mostly H_2S) on the P_d was studied by Carison and Cawston (Carlson and Cawston, 1996). Through total volume observations during the CCE test, a graphical model was proposed to predict the P_d through accurately determining the Z-factor. In sum, comparatively stress-free processing, ease of use and a general disregard of the temperature influence are the benefits and weaknesses of the experiential relationships (Nowroozi et al., 2009). Further, the reliance on primary derived data caused a decrease in appropriate EOS presentations upon application to new sites, and operators were requires to calibrate the related parameters (Jalili et al., 2007).

Thus, studies have yielded more beneficial, careful and appropriate approaches. Due to their inherent capacity for managing non-linearity, vagueness and uncertainty, scientists have used soft computing methods to overcome reservoir engineering problems, such as extracting PVT properties (Farasat et al., 2013 and Zendehboudi et al., 2012). For example, Akbari et al. applied a certain type of artificial neural network (ANN) to calculate the P_d using a group of thermodynamic and compositional features as the input (Jalili et al., 2007). Likewise, Nowroozi et al. constructed an adaptive neuro-fuzzy inference system (ANFIS) to predict the P_d by primarily considering compositional factors (Nowroozi et al., 2009). Similarly,Kaydani et al. (2013) proposed a conventional type of back-propagation artificial neural network to predict the P_d of lean retrograde gas condensate reservoirs.

For modern optimizing algorithms, it is valuable to refer to research by Rostami and Khakasr (2012), who proposed a model to predict the P_d through coupling Gaussian processes and particle swarm optimization methods. Moreover, this research was proposed as an easy-to-use, robust and sharp model for predicting dew point pressure (P_d) in retrograded gas condensate reservoirs. Thus, a hybrid composed of swarm intelligence and neural network as well as fuzzy logic (FL) coupled with genetic algorithm (GA)-fuzzy logic (FL) as robust artificial intelligent models was utilized to address the problems considered by this research. Thus, massive dew point pressure data banks extracted from previous studies (Ahmadi and Ebadi, 2014 and Al-Dhamen, 2010) were used in the aforementioned approaches for testing and validation. To confirm the capability and reliability of the evolved PSO-ANN approach, conventional correlations were used to estimate crude oil saturation pressure. The results for both intelligent and conventional correlations are shown in detail in the sections below. Additionally, descriptions of the models addressed herein are expanded in the following sections.

METHODOLOGY

Artificial Neural Network (ANN)

Artificial neural networks compose a bio-inspired approach, the initial pattern for which was determined from studying common human brain processes, that can numerically and inversely correlate relationships between input and output for each system due its distinct mathematical structure. The data are then technically implemented to train the network, and the network is used to estimate imprecise and vague data (Bain, 1873; James, 1890; Ahmadi and Shadizadeh, 2012; Ahmadi et al., 2013a; 2013b;Zendehboudi et al., 2013a, Zendehboudi et al., 2013b, Zendehboudi et al., 2012, Ahmadi, 2011, Ahmadi and Golshadi, 2012 and Ahmadi, 2012). The scheme depicted herein can be executed through relying on synchronous processing units, referred to as neurons and nodes, which are located in layers. The layers, input, output and hidden elements are the basic components of each artificial neural network (ANN), the number of neurons for which is determined based on the available data, designers and the goal of the study. Indisputably, the back-propagation feed forward network and multilayer perceptron (MLP) network, which are evaluated through classical techniques, require a significantly shorter development time, can use related information, and are the most favorable and common types of ANN in chemical engineering (Ahmadi and Shadizadeh, 2012, Ahmadi et al., 2014a, Ahmadi et al., 2014b, Ahmadi, 2014c, Ahmadi, 2014d and Ahmadi et al., 2013a; Zendehboudi et al., 2014, Zendehboudi et al., 2013a, Zendehboudi et al., 2013b, Ahmadi, 2011, Ahmadi and Golshadi, 2012, Ahmadi, 2012, Hagan et al., 1966, Vallés, 2006,Hornick et al., 1989, Hornik et al., 1990 and Garcia-Pedrajas et al., 2003).

Before detailing the main issue in this study, which is generating an up-to-the-minute optimization method out to precisely determine ANN-related parameters, including weights and biases, in terms of training, we also generated a conventional solution using the trial and error method. The database was divided into two main parts, which are referred to as the training and testing sets. The databases were divided to determine the most appropriate network structure by applying the larger group, the training data, while the test set, which was not previously used in the network during the training step, was used to examine the reliability of the proposed network for correlating saturation pressure. Optimization of the interconnected weights and node biases proceeded until the performance of the proposed ANN, which is based on statistical criteria, such as the mean squared error (MSE), is acceptable and when the output layer neuron output values are similar to the corresponding experimental data. The MSE is expressed as follows (Ahmadi and Shadizadeh, 2012, Ahmadi et al., 2014a, Ahmadi et al., 2014b, Ahmadi, 2014c, Ahmadi, 2014d, Ahmadi et al., 2013a, Zendehboudi et al., 2014, Zendehboudi et al., 2013a, Zendehboudi et al., 2013b, Ahmadi, 2011,Ahmadi and Golshadi, 2012, Ahmadi, 2012 and Hagan et al., 1966):

$$MSE^{Approach} = \frac{1}{2} \sum_{k=1}^{G} \sum_{j=1}^{m} [Y_j(k) - T_j(k)]^2$$

(1)

where the number of output nodes is indicated by m , the training samples are represented by G, and the expected and the actual outputs are denoted by $_{Yj}(k)$ and $_{Tj}(k)$, respectively. When the MSE gradually approaches zero, the error of our network model decreases (Ahmadi and Shadizadeh, 2012 and Ahmadi et al., 2013a; Zendehboudi et al., 2013a, Zendehboudi et al., 2013b, Zendehboudi et al., 2012, Ahmadi, 2011, Ahmadi and Golshadi, 2012 and Ahmadi, 2012).

Particle Swarm Optimization (PSO)

Particle swarm optimization (PSO) is a method that is mathematically inspired by studying and modeling the behavior of social organisms, such as a flock of birds. Similar to the GA, PSO is initiated using a population of random routs, which are referred to as particles; these particles should move within a defined search space at an adjustable velocity to determine the best position. Additionally, to maintain its focus on the target, each particle can update its velocity vector due to its flying experience and the flying experience of other particles in the search space, as illustrated in Fig. 1 (Ahmadi and Shadizadeh, 2012 and Ahmadi et al., 2013bb; Zendehboudi et al., 2013a, Zendehboudi et al., 2013b, Zendehboudi et al., 2012, Ahmadi and Golshadi, 2012 and Ahmadi, 2012; Ahmadi et al., 2014b).

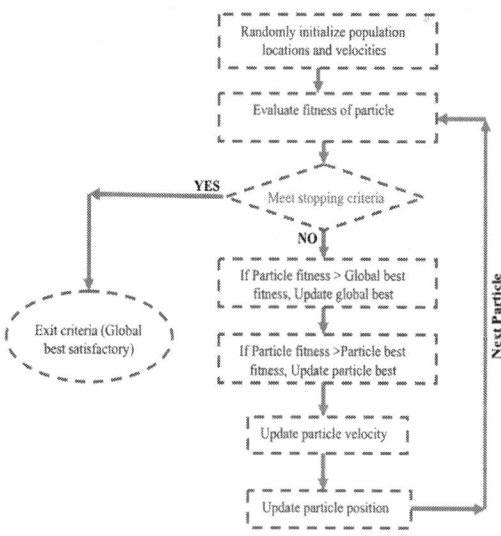

Figure 1: Flow chart for the particle swarm optimization process (Ahmadi and Shadizadeh, 2012, Zendehboudi et al., 2013a and Zendehboudi et al., 2013b).

RESULTS AND DISCUSSION

Fuzzy Logic Results

The results for the problem addressed herein yielded pdfs for all 14 attributes, the results for which are depicted in Fig. 2. Consequently, the center of values or pick for each pdf and the boundaries used to construct the MFs are shown in Fig. 3.

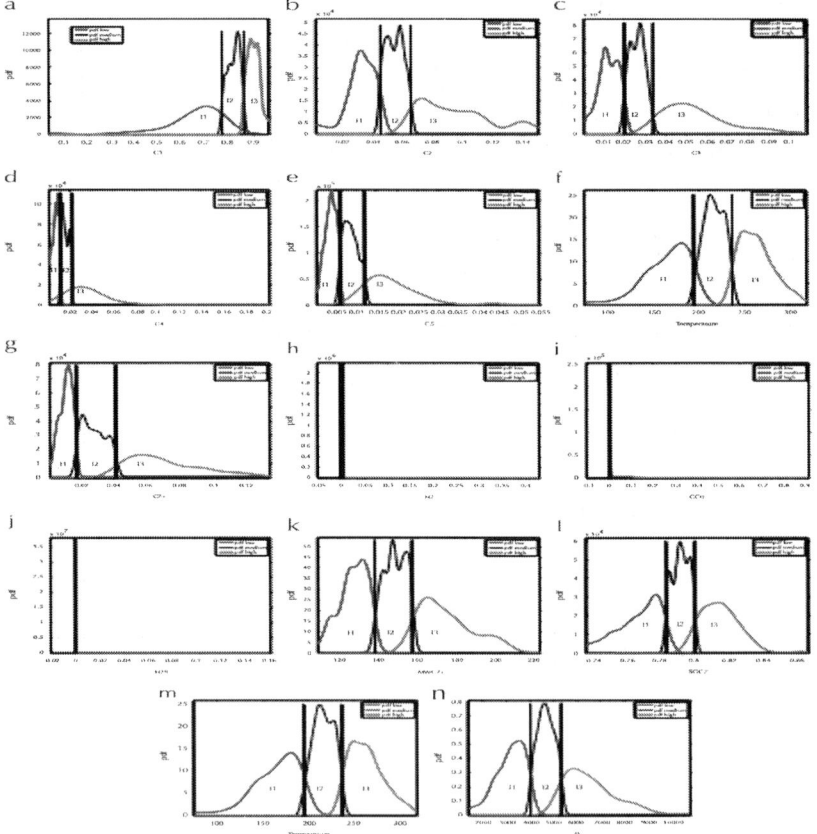

Figure 2: Estimated distribution related to intervals I1, I2, I3 for the following: a) C1, b) C2, c) C3, d) C4, e) C5, f) C6, g) C7+, h) N2, i) CO2, j) H$_2$S, k) MWC7+, l) SG C7+, m) T and n) P_d.

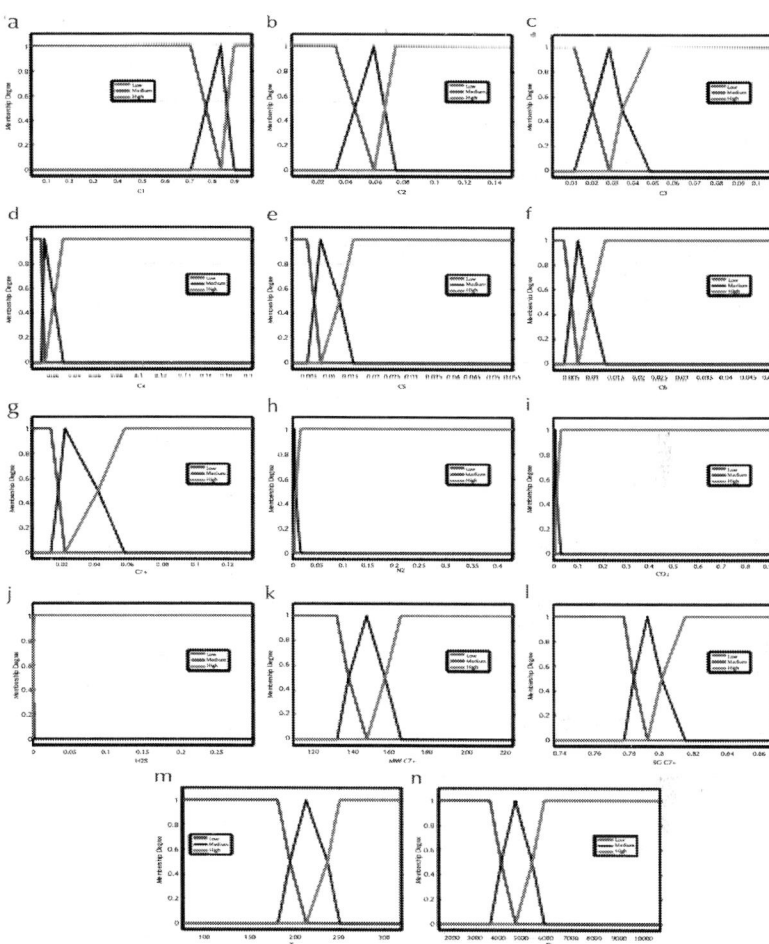

Figure 3: Membership functions for the decision classes L, M, and H connected with the following: a) C1, b) C2, c) C3, d) C4, e) C5, f) C6, g) C7+, h) N2, i) CO2, j) H_2S, k) MWC7+, l) SG C7+, m) T and n) P_d.

Fuzzy Logic and Genetic Algorithm Hybrid Results

The ANOVA assessment was used to discern the best rules and most influential parameters based on the data collected. (Fig. 4) The results show that the MW C7+ and SG C7+ can dramatically move the P_d, which is a strong function of the amount of T, C7+ and C1 and the as well. Additionally, P_d is infinitesimally impacted by parameters such as the mole percent of H_2S, N2, CO2, C6, C5, C4, C3 and C2.

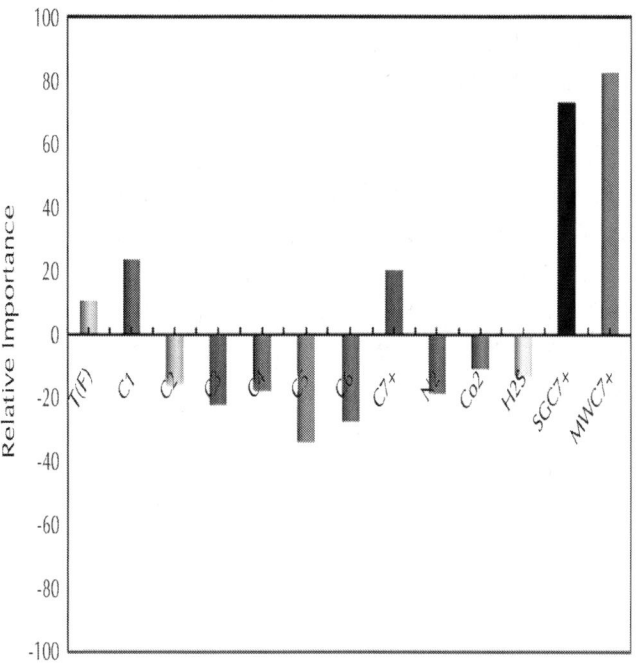

Figure 4: The ANOVA test results (Ahmadi and Ebadi, 2014).

Considering the constructed MFs and based on previous studies as well as comments by experienced experts, the following rules are proposed to predict the P_d.

- If *MW C7+* is high, and *SG C7+* is high, then P_d is high. (Weight$_1$=1).
- If *MW C7+* is medium, *SG C7+* is medium, *C1* is medium, *T* is medium, and *C7+* is medium, then P_d is medium (Weight$_2$=1).
- If *C2* is low, *C3* is low, *C4* is low, *C5* is low, *C6* is low, H_2S is low, *CO2* is low, and *N2* is low, then P_d is low. (Weight$_3$=1).

Each associated weight indicates the amount, effect and importance of the related rule compared with the others. The fuzzy rules introduced should tune the fuzzy interface system (FIS). To improve the performance of the FIS, the GA must be used to optimize the weights of the fuzzy rules and connected MFs parameters.

The GA with its special previously introduced operatives and characteristics must be used to extract the most optimized forms of the MFs and rules. Thus, the GA was set to minimize the MSE. To generate the most optimized general form of the FIS using the GA, a sensitivity analysis test must be performed to obtain the best arrangement of the GA. Additionally, the GA was initially set using the following features: crossover rate=2, mutation rate=0.0075, gene length=10, generation number=80 and initial population=8. The consequences of changing each of the operatives on the final, connected MSE for the most enhanced individual of the ultimate generation are shown in Fig. 5.

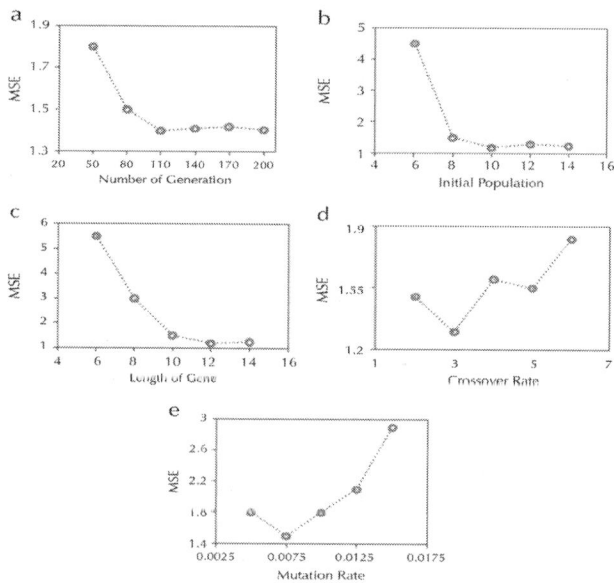

Figure 5: Sensitivity analysis using the GA parameters. a) Number of generations, b) initial population, c) gene length, d) crossover rate and e) mutation rate.

As shown, a GA with the features 110, 10, 12, 3 and 0.75 for the number of generations, initial population, gene length, crossover rate and mutation rate, respectively, can generate results with the highest performance. The few optimized GAs and previously developed FISs were used for the dataset, and the weights 78, 15, and 94 were used for rule 1, rule 2 and rule 3, respectively. Moreover, the MFs were optimized and are shown in Fig. 6.

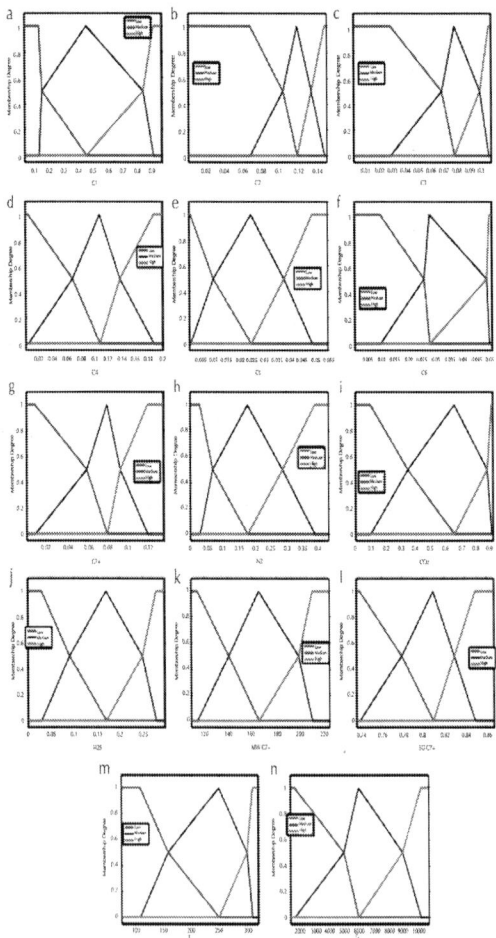

Figure 6: Membership functions for the decision classes L, M, and H connected with the following: a) C1, b) C2, c) C3, d) C4, e) C5, f) C6, g) C7+, h) N2, i) CO2, j) H_2S, k) MWC7+, l) SG C7+, m) T and n) P_d.

Eventually, the optimized FISs were used with the data, the optimization trend for which is plotted in Fig. 7; the results were graphed and are shown in Fig. 8.

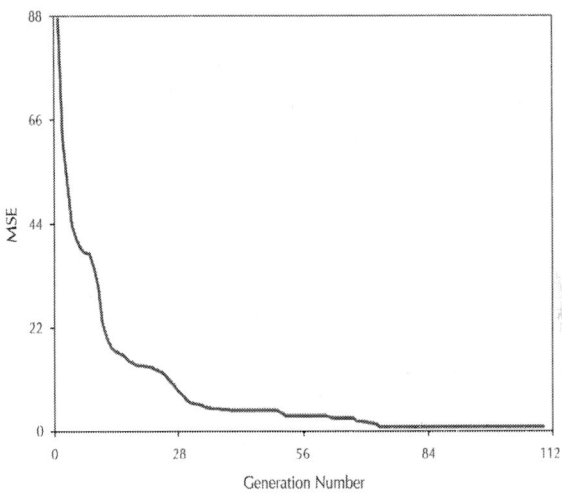

Figure 7: The optimization trend used by the FIS-GA.

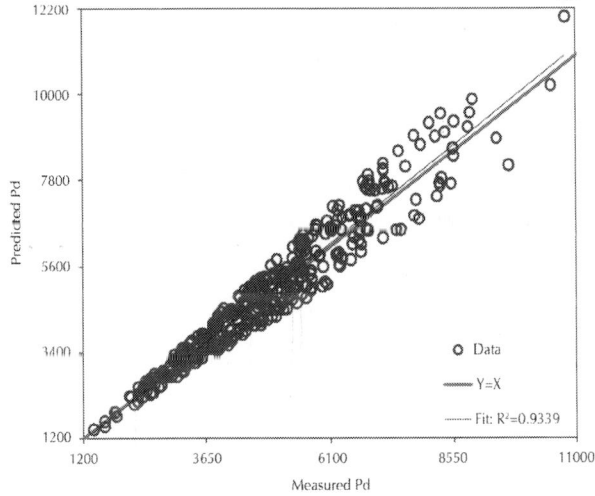

Figure 8: A scatter plot of the results compared with the measured values.

The best lines fitted to the data are described by the equation $y=1.0272x-92.54$. The relative errors of the values are illustrated based on the corresponding measured points in Fig. 9.

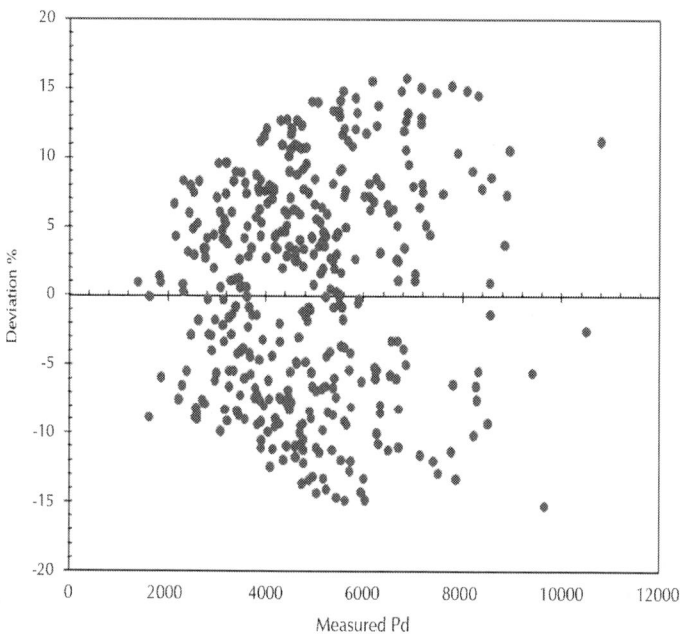

Figure 9: Relative error distribution versus the measured points.

Based on this figure, it can be inferred that the relative error distribution trend for the recovery factor data indicates maximum and minimum values near the middle and an improper peak at high pressures, but the data do not present a clear pattern in the plot of the net present value relative error distribution. In addition, the results were compared with the corresponding measured values based on their data indices, which are shown in Fig. 10.

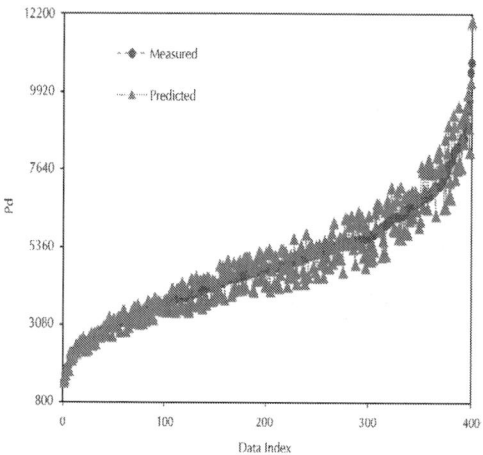

Figure 10: The results and measured values based on their corresponding indices.

As mentioned above, the statistical analysis shows that P_d is highly dependent on parameters such as MW C7+, SG C7+, T, C1 and C7+ to illustrate these situations; the results and corresponding measured values versus the relevant aforementioned parameters are shown in Fig. 11.

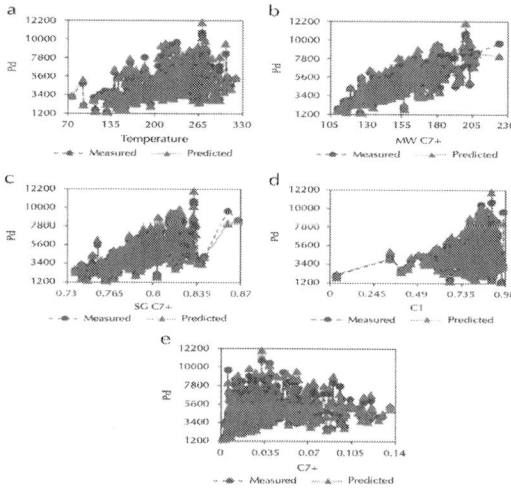

Figure 11: The results and measured corresponding values versus the most effective parameters. a) T, b) MW C7+, c) SG C7+, d) C1 and e) C7+.

Hybrid PSO and ANN Results

Fig. 12 shows a regression plot of the swarm intelligence output versus the corresponding real measured data. As depicted in Fig. 12, the output of the aforementioned model from both the testing and training phases follow a diagonal line (Y=X). In other words, the swarm model output is closest to the corresponding real values. Fig. 13 depicts the swarm model output and actual saturation pressure versus the corresponding data index. As illustrated in Fig. 13, the swarm model output follows the trend of the experimental saturation data. Finally, Fig. 14 shows the relative error distribution of the swarm model output versus the relevant dew point pressure data. Fig. 14 shows the maximum deviation of the aforementioned model output for the dew point pressure in the ranges 1500 to 2000 Psi, which is approximately 15%. Based on the relative deviation and correlation coefficient, which is referred to as robust statistical indexes, the evolved saturation pressure correlation is more rigorous than other approaches, such as the fuzzy logic and GA-fuzzy approaches, for estimating dew point pressure in retrograded gas condensate reservoirs. Finally, to consider previous outcomes, Table 1 shows details in terms of statistical indexes for the PSO-ANN method compared with other traditional solutions, such as Humoud and Al-Marhoun, Elsharkawy and Nementh and Kennedy. Based on Table 1, it can be concluded that the PSO-ANN is highly proficient and effective with a low vagueness, which are not characteristics of the models derived from the hybrid GA and fuzzy logic, artificial neural networks and other correlations, such as Humoud and Al-Marhoun as well as Elsharkawy, which yield high root mean square errors (RMSEs) and low correlation coefficients.

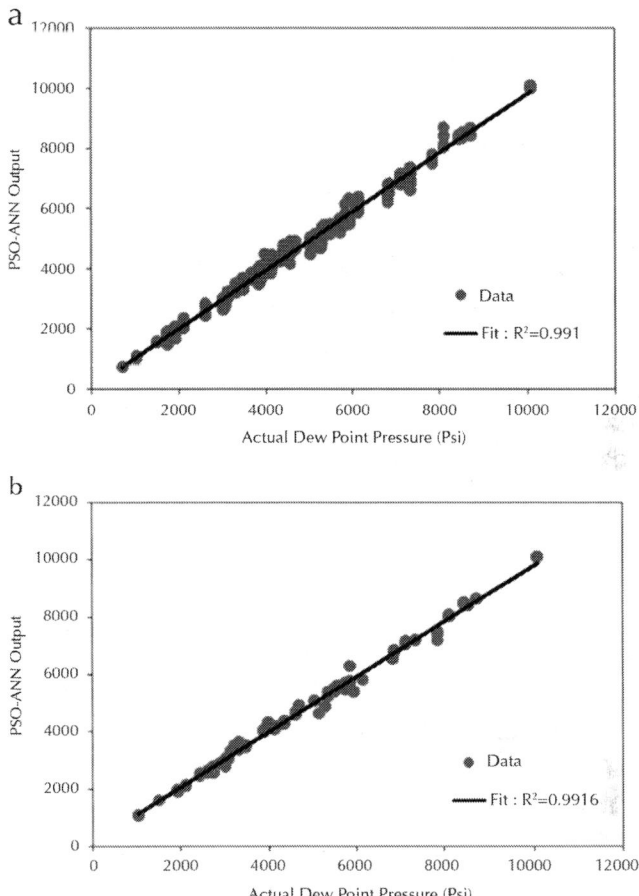

Figure 12: Regression plot of the PSO-ANN method for determining dew point pressure in retrograded gas reservoirs versus the relevant actual dew point pressure during the a) training phase and b) testing phase.

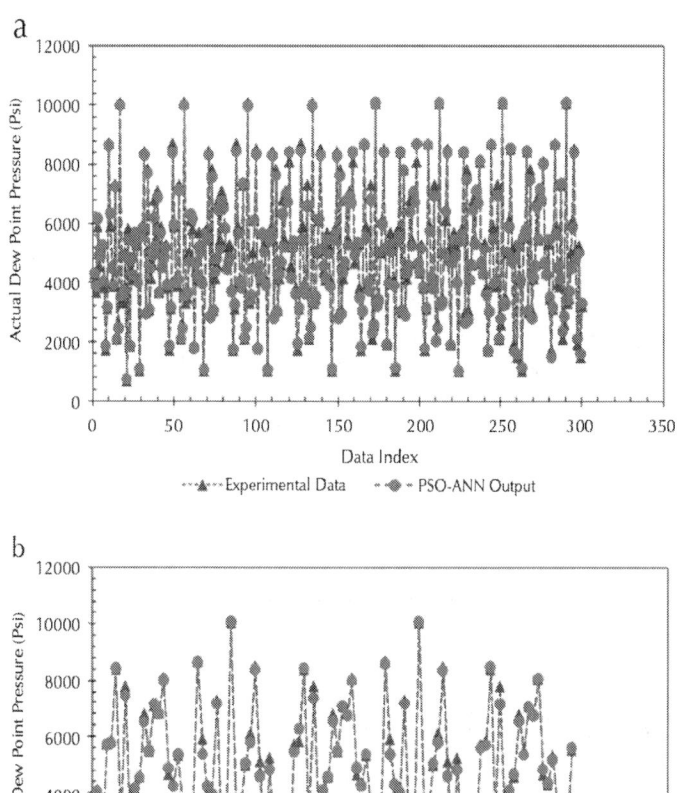

Figure 13: A parallel of the PSO-ANN method results and relevant actual dew point pressure against corresponding data index during the a) training phase and b) testing phase.

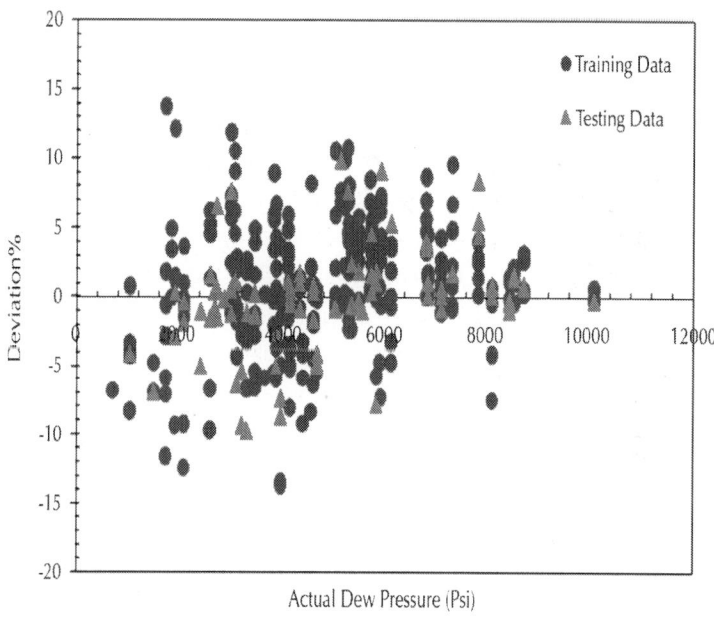

Figure 14: Error distribution for the PSO-ANN output versus the relevant actual dew point pressure.

CONCLUSIONS

An accurate determination of the dew point pressure (P_d) through retrograded condensate gas reservoirs has a strong impact on simulating gas/condensed gas flow through a porous medium. Therefore, in this research, much effort was expended to estimate the dew point pressure (P_d) with high accuracy and low uncertainty. Therefore, swarm intelligence and neural networks were used to develop an efficient approach that solves the aforementioned problem with acceptable precision and accuracy. Furthermore, the accurate experimental dew point pressure (P_b) data reported in previous surveys were introduced into the proposed model to tune and validate the suggested approach. Based on this research, the following is deduced.

- Acceptable similarity between the dew point pressures (P_d) obtained from the suggested swarm intelligence model compared with the relevant actual dew point pressure (P_d) values was observed. In other words, previous models have failed to estimate the dew point pressure (P_d) based on the calculated statistical parameters of each aforementioned model.

- The proposed smart model (PSO-ANN) to determine the dew point pressure (P_d) through retrograded gas condensate reservoirs is easy-to-use, is cheap and performs at a high level. Furthermore, it is beneficial and user friendly for tuning the integrity and performance of marketable reservoir simulators, such as PVTi in the ECLIPSE package, and for matching the history and other aspects of condensate production with the model developed herein. Moreover, the proposed smart model can be used as a substitute when essential dew point pressure data are unavailable.

REFERENCES

1.   Ahmadi, M.A., Ebadi, M., Shokrollahi, A., Majidi, S.M.J., 2013b. Evolving artificial neural network and imperialist competitive algorithm for prediction oil flow rate of the reservoir. Appl. Soft Comput. 13 (2), 1085–1098.

2.   Ahmadi, M.A., Shadizadeh, S.R., 2012. New approach for prediction of asphaltene precipitation due to natural depletion by using evolutionary algorithm concept. Fuel 102 (0), 716–723.

3.   Ahmadi, M.A., Ebadi, M., Hosseini, S.M., 2014a. Prediction breakthrough time of water coning in the fractured reservoirs by implementing low parameter support vector machine approach. Fuel 117 (Part A), 579–589.

4.   Ahmadi, M.A., 2012. "Neural Network Based Unified Particle Swarm Optimization for Prediction of Asphaltene Precipitation". Fluid Phase Equilibria 314, 46–51.

5.   Ahmadi, M.A., 2011. "Prediction of asphaltene precipitation using artificial neural network optimized by imperialist competitive algorithm". J. of Petroleum Exploration and Production Technology 1, 99–106.

6.   Ahmadi, M.A., Ebadi, M., 2014. Evolving Smart Approach for Determination Dew Point Pressure of Condensate Gas Reservoirs. J. of Fuel, Fuel 117 (Part B), 1074–1084.

7.   Ahmadi, M.A., Golshadi, M., 2012. "Neural Network Based Swarm Concept for Prediction Asphaltene Precipitation due Natural Depletion". J. of Petroleum Science and Engineering 98–99, 40–49.

8.   Ahmadi, M.A., Zendehboudi, S., Lohi, A., Elkamel, A., Chatzis, I., 2013a. "Reservoir permeability prediction by neural networks combined with hybrid genetic algorithm and particle swarm optimization ". Geophysical Prospecting 61, 582–598.

9.   Ahmadi, M.A., Masoumi, M., Kharrat, R., Mohammadi, A.H., 2014b. Gas Analysis by In Situ Combustion in Heavy Oil Recovery Process: Experimental and Modeling Studies. J. Chemical Eng. Tech. 37 (3), 1–11.

10. Ahmadi, M.A., Ebadi, M., Marghmaleki, P.S., Fouladi, M.M., Evolving Predictive Model to Determine Condensate-to-Gas Ratio in Retrograded Condensate Gas Reservoirs, Fuel (2014c), 124C, pp. 241-257.

11. Ahmadi, M.A., Ebadi, M., Samadi, A., Zendedel-Siuki, M., Phase Equilibrium Modeling of Clathrate Hydrates of Carbon Dioxide + 1,4-Dioxine Using Intelligent

12. Approaches, J. Dispe. Sci. Tech. 2014d, http://dx.doi.org/10.1080/01932691.2014. 904792.

13. Alavi, F.S., Mowla, D., Esmaeilzadeh, F., 2010. Production performance analysis of Sarkhoon gas condensate reservoir. J. of Petroleum Science and Engineering 75, 44–53.

14. Al-Dhamen, M.A., New Models for Estimating the Dew-point Pressure of Gas Condensate Reservoir, King Fahd University of Petroleum & Minerals, 2010.

15. Alves, J.C.L., Henriques, C.B., Poppi, R.J., 2012. Determination of diesel quality parameters using support vector regression and near infrared spectroscopy for an in-line blending optimizer system. Fuel 97, 710–717.

16. App, J.F., Burger, J.E., 2009. Experimental Determination of Relative Permeabilities for a Rich Gas/Condensate System Using Live Fluid. SPE Reserv. Evaluation & Engineering 12, 263–269.

17. Bain, A., 1873. Mind and Body: The Theories of Their Relation. D. Appleton and Company, New York.

18. Bashipour, F., Ghoreishi, S.M., 2012. Experimental optimization of supercritical extraction of β-carotene from Aloe barbadensis Miller via genetic algorithm. The J. of Supercrit. Fluids 72, 312–319.

19. Bayat, M., Dehghani, Z., Hamidi, M. & Rahimpour, M.R. Methanol synthesis via sorption-enhanced reaction process: Modeling and multi-objective optimization. Journal of the Taiwan Institute of Chemical Engineers.

20. Berning, T., 2012. The dew point temperature as a criterion for optimizing the operating conditions of proton exchange membrane fuel cells. International J. of Hydrog. Energy 37, 10265–10275.

21. Beynon, M.J., Peel, M.J., Tang, Y.-C., 2004. The application of fuzzy decision tree analysis in an exposition of the antecedents of audit fees. Omega 32, 231–244.

22. Bonyadi, M., Esmaeilzadeh, F., 2007. Prediction of gas condensate properties by Esmaeilzadeh–Roshanfekr equation of state. Fluid Phase Equilibria 260, 326–334.

23. Brown, A.S., Milton, M.J.T., Vargha, G.M., Mounce, R., Cowper, C.J., Stokes, A.M.V.,

24. Benton, A.J., Lander, D.F., Ridge, A., Laughton, A.P., 2009. Measurement of the Hydrocarbon Dew Point of Real and Synthetic Natural Gas Mixtures by Direct and Indirect Methods. Energy & Fuels 23, 1640–1650.

25. Carlson, M.R. & Cawston, W.B. 1996. Obtaining PVT Data For Very Sour Retrograde Condensate Gas and Volatile Oil Reservoirs: A Multi-disciplinary Approach. SPE Gas Technology Symposium. Calgary, Alberta, Canada.

26. Chowdhury, N.S., Sharma, R., Pope, G.A., Sepehrnoori, K., 2008. A Semianalytical Method To Predict Well Deliverability in Gas-Condensate Reservoirs. SPE Reserv. Evaluation & Engineering 11, 175–185.

27. Da Silva, M.P.F., Brito, L.R.E., Honarato, F.A., Paim, A.P.S., Pasquini, C., Pimentel, M.F., 2014. Classification of gasoline as with or without dispersant and detergent additives using infrared spectroscopy and multivariate classification. Fuel 116, 151–157.

28. El-Ssebakhy, E.A., Asparouhv, O., Abdulraheem, A.-A., Al-Majed, A.-A., Wu, D., Latinski, K., Raharja, I., 2012. Functional networks as a new data mining predictive paradigm to predict permeability in a carbonate reservoir. Expert. Systems with Applications 39, 10359–10375.

29. Elsharkawy, A.M., Foda, S.G., 1998. EOS Simulation and GRNN Modeling of the Constant Volume Depletion Behavior of Gas Condensate Reservoirs. Energy & Fuels 12, 353–364

30. Elsharkawy, A.M., 2002a. Predicting the dew point pressure for gas condensate reservoirs: empirical models and equations of state. Fluid Phase Equilibria 193, 147–165.

31. Elsharkawy, A., 2002b. Predicting the dewpoint pressure for gas condensate reservoir: empirical models and equations of state. Fluid Phase Equilib 193, 147–165.

32. Farasat, A., Shokrollahi, A., Arabloo, M., Gharagheizi, F., Mohammadi, A.H., 2013. Toward an intelligent approach for determination of saturation pressure of crude oil. Fuel Processing Technology 115, 201–214.

33. Fu, G., 2008. A fuzzy optimization method for multicriteria decision making: An application to reservoir flood control operation. Expert. Systems with Applications 34, 145–149.

34. Garcia-Pedrajas, N., Hervas-Martinez, C., Munoz-Perez, J., 2003. COVNET: A cooperative co evolutionary model for evolving artificial neural networks. IEEE Transaction on Neural Networks, 575–59614, 575–596.

35. Ghiasi-Freez, J., Kadkhodaie-Ilkhchi, A., Ziaii, M., 2012. Improving the accuracy of flow units prediction through two committee machine models: An example from the South Pars Gas Field, Persian Gulf Basin, Iran. Computers & Geosciences 46, 10–23.

36. Hagan, M.T., Demuth, H.B., Beal, M., 1966. Neural Network Design. PWS Publishing Company, Boston.

37. Hornick, K., Stinchcombe, M., White, H., 1989. Multilayer feed forward networks are universal approximators. Neural Networks (2), 359–366.

38. Hornik, K., Stinchcombe, M., White, H., 1990. Universal approximation of an unknown mapping and its derivatives using multilayer feed forward networks. Neural Networks 3 (5), 551–600.

39. Humoud, A.A., Al-Marhoun, M.A., A new correlation for gas condensate dew point pressure prediction. Paper SPE 68230 Presented at the 2001 SPE Middle East Oil Show, Bahrain, 17–20 March.

40. Jafari Kenari, S.A., Mashohor, S., 2013. Robust committee machine for water saturation prediction. J. of Petroleum Science and Engineering 104, 1–10.

41. Jalili, F., Abdy, Y. & Akbari, M.K. 2007. Dewpoint Pressure Estimation of Gas Condensate Reservoirs, Using Artificial Neural Network (ANN). EUROPEC/EAGE Conference and Exhibition. London, U.K.

42. James, W., 1890. The Principles of Psychology. H. Holt and Company, New York. Nemeth, L.K., Kennedy, H.T., 1967. A Correlation of Dewpoint Pressure With Fluid Composition and Temperature. SPE J 7, 99–104.

43. Kaydani, K., Haghizade, A., Mohebbi, A., 2013. A Dew Point Pressure Model for Gas Condensate Reservoirs Based

on anArtificial Neural Network. Petroleum Science and Technology 31, 1228–1237.

44. Li, C., Jia, W., Wu, X., 2012. Application of Lee-Kesler equation of state to calculating compressibility factors of high pressure condensate gas. Energy Procedia 14, 115–120.

45. Louli, V., Pappa, G., Boukouvalas, C., Skouras, S., Solbraa, E., Christensen, K.O., Voutsas, E., 2012. Measurement and prediction of dew point curves of natural gas mixtures. Fluid Phase Equilibria 334, 1–9.

46. Luo, K., Li, S., Zheng, X., Chen, G., Liu, N. & Sun, W. 2001. Experimental Investigation Into Revaporization of Retrograde Condensate. SPE Production and Operations Symposium. Oklahoma City, Oklahoma.

47. Marruffo, I., Maita, J., Him, J., Rojas, G., "Correlation To Determine Retrograde Dew Pressure and C7þ Percentage of Gas Condensate reservoirs on Basis of Production test Data of Eastern Venezuelan Fields" presented at the SPE Gas Technology Symposium held in Calgary, Alberta, Canada, 30 April 2002.

48. Marruffo, I., Maita, J., Him, J. & Rojas, G. 2001. Statistical Forecast Models To Determine Retrograde Dew Pressure and C7þ Percentage of Gas Condensates on Basis of Production Test Data of Eastern Venezuelan Reservoirs. SPE Latin American and Caribbean Petroleum Engineering Conference. Buenos Aires, Argentina.

49. Martins, M.S.R., Fuchs, S.C., Pando, L.U., Luders, R., Delgado, M.R., 2013. PSO with path relinking for resource allocation using simulation optimization. Computers & Industrial Engineering 65, 322–330.

50. Mohammadi, H., Sedaghat, M.H., Khaksar Manshad, A., 2013. Parametric investigation of well testing analysis in low permeability gas condensate reservoirs. J. of Nat. Gas Science and Engineering 14, 17–28.

51. Mokhtari, R., Varzandeh, F., Rahimpour, M.R., 2013. Well productivity in an Iranian gas-condensate reservoir: A case study. J. of Nat. Gas Science and Engineering 14, 66–76.

52. Nasrifar, K., Bolland, O., Moshfeghian, M., 2005. Predicting Natural Gas Dew Points from 15 Equations of State. Energy & Fuels 19, 561–572.

53. Nejad Ebrahimi, A., Jamshidi, S., Iglauer, S., Boozarjomehri, R.B., 2013. Genetic algorithm-based pore network extraction from micro-computed tomography images. Chemical Engineering Science 92, 157–166.

54. Nemeth, L.K., Kennedy, H.T., "A Correlation of Dewpoint Pressure with Fluid Composition and Temperature," paper SPE 1477 presented at SPE 41st Annual Fall Meeting held in Dallas, Texas, 1966.

55. Nowroozi, S., Ranjbar, M., Hashemipour, H., Schaffie, M., 2009. Development of a neural fuzzy system for advanced prediction of dew point pressure in gas condensate reservoirs. Fuel Processing Technology 90, 452–457.

56. Olatunji, S.O., Selamat, A., Abdul Raheem, A.A. Improved sensitivity based linear learning method for permeability prediction of carbonate reservoir using interval type-2 fuzzy logic system. Applied Soft Computing.

57. Rahimpour, M.R., Seifi, M., Paymooni, K., Shariati, A., Raeissi, S., 2011. Enhancement in NGL production and improvement in water dew point temperature by optimization of slug catchers' pressures in water dew point adjustment unit. J. of Nat. Gas Science and Engineering 3, 326–333.

58. Rostami, H., Khakasr, A., 2012. Application of Evolutionary Gaussian Processes Regression by Particle Swarm Optimization for Prediction of Dew Point Pressure in Gas Condensate Reservoirs. Neural Computing and Applications, 1–9.

59. Sadeghi Booghar, A., Masihi, M., 2010. New technique for calculation of well deliverability in gas condensate reservoirs. J. of Nat. Gas Science and Engineering 2, 29–35.

60. Salamchi, A., Sayyafzadeh, M., Haghghi, M., 2013. Infill well placement optimization in coal bed methane reservoirs using genetic algorithm. Fuel 111, 248–258.

61. Shadizadeh, S.R., Rashtchian, D. & Moradi, S. 2006. Simulation of Experimental GasRecycling Experiments in Fractured Gas/Condensate Reservoirs. SPE Gas Technology Symposium. Calgary, Alberta, Canada.

62. Shen, P., Luo, K., Zheng, X., LI, S., Dai, Z. & Liu, H. 2001. Experimental Study of NearCritical Behavior of Gas Condensate Systems. SPE Production and Operations Symposium. Oklahoma City, Oklahoma.

63. Soleimani, R., Shoushtari, N.A., Mirza, B., Salahi, A., 2013. Experimental investigation, modeling and optimization of membrane separation using artificial neural network and multi-objective optimization using genetic algorithm. Chemical Engineering Res. and Des 91, 883–903.

64. Thomas, F.B., Bennion, D.B., Andersen, G., 2009. Gas Condensate Reservoir Performance. J. of Canadian Petroleum Technology 48, 18–24.

65. Ursin, J.-R., 2004. Fluid flow in gas condensate reservoirs: the interplay of forces and their relative strengths. J. of Petroleum Science and Engineering 41, 253–267.

66. Vallés, H.R. A Neural Networks Method to Predict Activity Coefficients for Binary Systems Based on Molecular Functional Group Contribution. Master thesis, University of Puerto Rico, 2006.

67. Wali, W.A., Al-Shamma›A, A.I., Hassan, K.H., Cullen, J.D., 2012. Online genetic-ANFIS temperature control for advanced microwave biodiesel reactor. J. of Process. Control 22, 1256–1272.

68. Wang, L., Yang, Z., He, M., 2012. Research on optimizing control model of hydrogen fueled engines based on thermodynamics and state space analysis method about nonlinear system. International J. of Hydrog. Energy 37, 9902–9913.

69. Yong, L., Baozhu, L., Yongle, H., Yuwei, J., Weihong, Z., Xiangjiao, X., Yu, N., 2010.

70. Water production analysis and reservoir simulation of the Jilake gas condensatefield. Petroleum Exploration and Development 37, 89–93.

71. Yu, S., Guo, X., Zhu, K., Du, J., 2010. A neuro-fuzzy GA-BP method of seismic reservoir fuzzy rules extraction. Expert. Systems with Applications 37, 2037–2042.

72. Zahraie, B., Hosseini, S.M., 2009. Development of reservoir operation policies considering variable agricultural water demands. Expert. Systems with Applications 36, 4980–4987.

73. Zendehboudi, S., Ahmadi, M.A., Bahadori, A., Shafiei, A., Babadagli, T., 2013a. A Developed Smart Technique to Predict Minimum Miscible Pressure—EOR Implication. The Canadian J. of Chemical Engineering 91 (7), 1325–1337.

74. Zendehboudi, S., Ahmadi, M.A., Mohammadzadeh, O., Bahadori, A., Chatzis, I., 2013b. Thermodynamic Investigation of Asphaltene Precipitation during Primary Oil Production: Laboratory and Smart Technique. Ind. Eng. Chem. Res 52, 6009–6031.

75. Zendehboudi, S., Ahmadi, M.A., James, L., Chatzis, I., 2012. Prediction of Condensateto-Gas

76. Ratio for Retrograde Gas Condensate Reservoirs Using Artificial Neural Network with Particle Swarm Optimization. Energy & Fuels 26, 3432–3447.

77. Zendehboudi, S., Ahmadi, M.A., Bahadori, A., Lohi, A., Elkamel, A., Chatzis, I., 2014. Estimation of Breakthrough Time for Water Coning in Fractured Systems: Experimental Study and Connectionist Modeling. AICHE J. 60 (5), 1905–1919.

78. Zheng, X., Sheng, P., LI, S., Luo, K. & Dai, Z. 2000. Experimental Investigation into Near-Critical Phenomena of Rich Gas Condensate Systems. International Oil and Gas Conference and Exhibition in China. Beijing, China.

Reservoir Geochemistry – A Reservoir Engineering Perspective

W.A. England

BP Exploration, Chertsey Road, Sunbury-on-Thames, Middlesex, TW16 7LN, UK

ABSTRACT

This paper reviews the applications of reservoir geochemistry from a reservoir engineering point of view. Some of the main tasks of reservoir engineering are discussed with an emphasis on the importance of appraising reservoirs in the pre-development stage. A brief review of the principal methods and applications of reservoir geochemistry are given, in the context of applications to reservoir engineering problems. The importance of compositional differences in fluid samples from different depths or spatial locations is discussed in connection with the identification of internal flow barriers. The importance of understanding the magnitude and origin of vertical compositional gradients is emphasised because of possible confusion with purely lateral changes. The geochemical origin and rate of dissipation of compositional differences over geological time is discussed.

Geochemical techniques suitable for bulk petroleum fluid samples include GC fingerprinting, GCMS, isotopic and PVT measurements. Core sample petroleum extracts may also be studied by standard geochemical methods but with the added complication of possible contamination by drilling mud. Aqueous phase residual salt extracts can be studied by strontium isotope analysis from core samples. Petroleum fluid inclusions allow the possibility of establishing the composition of paleo-accumulations.

The problems in predicting flow barriers from geochemical measurements are discussed in terms of "false positives" and "false negatives". Suggestions are made for areas that need further development in order to encourage the wider acceptance and application of reservoir geochemistry by the reservoir engineering community. The importance of integrating all available data is emphasised.

Reservoir geochemistry may be applied to a range of practical engineering problems including production allocation, reservoir compartmentalisation, and the prediction of gravitational gradients. In this review applications where reservoir geochemistry is being used successfully are contrasted with those where applications are less well established.

INTRODUCTION

Reservoir geochemistry is concerned with establishing the spatial and temporal chemical variation of petroleum and aqueous phases in oil and gas reservoirs. The goal is to understand these variations in terms of how accumulations filled from an evolving source rock system, together with the effects of processes such as diffusive and convective mixing, gravitational and thermal segregation, biodegradation, phase changes and leakage. An important practical aim is to use this information to distinguish regions of poor communication in reservoirs because this has a large impact on oil and gas production.

Whereas reservoir geochemistry studies processes occurring over thousands or millions of years, reservoir engineering operates over timescales of no more than years or tens of years. Reservoir engineering is a vast subject, well covered by a number of text books such as that by Dake (1985). Reservoir engineering applies physical and chemical principles to plan and implement the optimal development of petroleum reservoirs. It is involved in all stages of field life from discovery and appraisal to production and abandonment.

Since its growth over the last 20 yr, reservoir geochemistry and its associated techniques have been extensively reviewed elsewhere (e.g. Larter and Aplin, 1994, Cubitt and England, 1995, Larter et al., 1995, Peters and Fowler, 2002 and Cubitt et al., 2004). This paper therefore concerns itself mainly with a reservoir engineering perspective on reservoir geochemistry. Westrich et al. (1999) p. 514 remarked that:

"Despite the applicability and promise of this new technology, its true value is not widely recognised in the industry. An important reason is that a variety of issues and problems need to be addressed to increase our level of understandin .We need to know when it works, when it does not and why".

In this review I discuss the areas where reservoir geochemistry already is being used successfully and contrast these with areas where less progress has been made. I also suggest areas for further research and development aimed at extending the reliability and applicability of reservoir geochemistry to a wider range of reservoir engineering problems. The following sections discuss the major reservoir engineering activities during the appraisal and producing phases of field life, with the aim of putting reservoir geochemical applications in context.

APPRAISAL

Following a successful exploration well, appraisal is aimed at reducing the key uncertainties in a proposed oil or gas field development to the point where a decision to develop can proceed with reasonable confidence. Considerable sums of money are involved: placing a new oil or gas project on stream requires a major investment — it may be necessary to expend as much as $1 billion to establish a new field.

During the appraisal of newly discovered resources a number of engineering options will be compared for producing oil and gas from a particular field. These options include the need for water or gas injection (to maintain pressure and improve recovery), the number, type and location of production and injection wells and the prediction of annual oil, gas and water profiles for economic evaluation. The accuracy of these oil and gas production profiles is very important because they are used to design the size and scope of the expensive production facilities — any departures from plan may lead to production restrictions or expensive modifications to existing equipment.

A crucial point is that the majority of the capital for many fields is spent in advance of the availability of sustained production performance data. It is often not possible to be certain how well (or if) the various geologically defined intervals and fault segments comprising the reservoir are connected until the wells have flowed a significant quantity of petroleum to surface. This means that many crucial decisions are made on the basis of very limited data. Because many development wells are drilled before production has started, there is a significant risk of poor well positioning. Appraisal is therefore the stage at which reservoir geochemistry has the greatest impact on investment decisions, particularly those involving reservoir connectivity and reservoir compartmentalisation.

Deciding the Most Important Topics to Study during Appraisal

An important planning tool for appraisal studies is the sensitivity study. This shows the sensitivity of key objectives (such as oil reserves, production rate, and net present value) to the uncertain geological and engineering description of the reservoir undergoing appraisal. An example was given by Smith et al. (1993) from an offshore field (UKCS) when it was at the end of the appraisal phase. They showed that the most important five parameters influencing uncertainty in oil reserves were:

- Oil initially in place
- The presence or absence of oil-charged sand in the undrilled "Western Flank"
- Residual oil saturation
- Oil–water relative permeability
- Absolute permeability

In this example, it is somewhat difficult to envisage reservoir geochemistry reducing reserves uncertainty during late appraisal. However, if geochemical data had been available from the Western Flank following an additional appraisal well, reservoir geochemistry could have been used to assess its degree of connectivity to the rest of the field.

Of course, every reservoir has a different set of key parameters contributing to uncertainty. In many cases, reservoir geochemistry makes important contributions, especially in the area of fluid properties and reservoir compartmentalisation as described below.

The important point is that each appraisal problem should be analysed by a sensitivity study to ensure that work is appropriately focussed. The applicability of reservoir geochemistry should be assessed regularly by considering whether geochemical studies can reduce the uncertainties in any of the key parameters.

However it is important that the application of reservoir geochemical tools is considered, so that a rational decision is made as to its application. Reservoir geochemistry can then be assessed in relation to the many other appraisal techniques such as new well drilling, geophysical work, petrophysics, sedimentology, structural geology and rock mechanics. The following sections describe some of the areas where reservoir geochemistry has been applied to appraisal problems.

Measurements on Petroleum Fluids

Petroleum fluids are at the heart of any appraisal programme. Measurements carried out for engineering purposes – sometimes known as "PVT" (pressure, volume and temperature) measurements – include viscosity, gas oil ratio (GOR) and formation volume factor (FVF). FVF is defined as the subsurface volume of reservoir fluid required to produce 1 m^3 of stabilised stock tank oil (units m^3/m^3). These measurements are used to choose the depletion scheme, and to design the wells, flow-lines and fluid processing facilities. Non hydrocarbon components (such as CO_2, H_2S, mercaptans, Hg) need to be studied, as they have an important impact in terms of the extra processing needed to remove them or the possible additional cost of anti-corrosion steel. McCain (1990) gives a fuller discussion of petroleum fluid properties and their petroleum engineering impacts.

In many reservoirs, viscosity is one of the key uncertainties. Not only does viscosity determine the rate at which oil can flow into the production wells, but it also affects the efficiency of the displacement of oil by injected gas or water by changing the fractional flow curve. In complex reservoirs, viscosity may vary vertically and/or laterally across a field, either within or between different intervals and/or fault compartments, Smalley et al. (1997). Since it will not be possible to sample and measure viscosity at all possible development well locations, reservoir geochemistry can play a vital role in improving viscosity predictions during the appraisal of new discoveries.

Additional "assay" measurements are made on stock tank oil to assess its market value to oil refiners — important parameters include distillation cuts, % sulphur, total acid number, trace metals and mercaptan content. The prediction of these properties in non-drilled segments is an important activity that should involve reservoir geochemical considerations. The asphaltene, wax and hydrate contents of oils, and their possible variations also need to be assessed to ensure that they do not cause blockages in wells or pipelines. Again, this is an area where reservoir geochemistry can be used to predict the properties of fluids at undrilled locations.

It is unusual for geochemical measurements (such as GC fingerprints, isotope and biomarker ratios) to be of any direct interest to reservoir engineers. However, the interpretations that can be made from geochemical measurements are of crucial interest and have a large impact when translated into cost and production forecasts by the reservoir and facility engineers.

Vertical Gradients and their Origin

Correctly appraising the vertical gradient in a petroleum column is an important first step in building a 3D assessment of the fluids in a petroleum discovery. This is used to calculate oil and gas reserves from parameters such as GOR and FVF. There are important implications for facility design: for example if GOR is poorly estimated, then production may need to be restricted if insufficient gas processing is available on the installed equipment.

It has been long known that a static column of petroleum in a gravitational field evolves towards a thermodynamic equilibrium state in which GOR increases towards the top of the column and denser components tend to concentrate towards the base. Most PVT and other properties show a continuous, smooth variation with depth under these conditions. For fuller details see Sage and Lacey (1939) and a recent review by Høier and Whitson (2001).

Ab initio predictions of gravitational gradients are very difficult. Attempts using equation of state methods are often unsuccessful unless the parameters have been adjusted (or "tuned") to reproduce the observed depth trends, Ratulowski et al. (2000). The observed range of natural vertical gradients is also wide, and the controlling parameters are not understood in full. High asphaltene content (Hirchberg, 1988), high aromatic content (Schulte, 1980) and proximity to the fluid's critical point, Høier and Whitson (2001), have been discussed as important in the development of large vertical gradients.

Ratulowski et al. (2000) showed that the sizable PVT differences observed between fluids from different appraisal wells in Bullwinkle Field (Gulf of Mexico) were entirely consistent with vertical gravitational segregation. Equation of state predictions tuned to PVT data could not initially account for the differences seen between wells. However, an experiment in which a fluid sample was centrifuged in the laboratory was able to very closely reproduce the observed Bullwinkle vertical PVT variations. This removed a possible interpretation of the Bullwinkle PVT data that would have implied compartmentalisation, greatly affecting the number of wells needed to drain the field. Ratulowski et al. (2000) remarked that it was not surprising the equation of state models were poor at predicting gravitational gradients as they were tuned on laboratory separator data that does not involve gravity effects. It is hoped that in the future sufficient centrifuge data will be available to remedy this situation.

The rate at which the gravitational gradient is set up by diffusion is relatively fast on a geological timescale, England et al. (1987) —

$$\text{Diffusive Timescale} \sim 0.1L^2/D \qquad (1)$$

Where L (m) is the distance between the two points over which diffusion is occurring and D is the diffusion constant (m²/s).

Ratulowski et al. (2000) showed that in a dipping sheet reservoir, coupled diffusion and convection is faster than that estimated for diffusion alone from Eq. (1) since diffusion only needs to occur vertically through the stratigraphic interval, setting up a density instability that is eliminated by a series of convection cells. (This was confirmed by their laboratory data).

Some authors have proposed that thermally induced gradients could modify the isothermal gravitationally induced compositional gradient. These are discussed by Høier and Whitson (2001). One problem in their implementation is a lack of laboratory data to confirm or invalidate the various models proposed. Field data are of course always open to many interpretations.

In the presence of a sufficiently high vertical temperature gradient, spontaneous convection can occur, which tends to eliminate any gravitational gradient — although there is an intermediate thermal-gravitational regime that has been discussed by Jacqmin (1990) and Ghorayeb and Firoozabadi (2000).

Other causes of vertical compositional differences need to be considered. Very recent filling, or currently occurring water-washing or biodegradation can be invoked to provide alternative explanations for vertical compositional gradients. In this case geochemical measurements on the petroleum samples can provide useful discrimination between models. However, because diffusive mixing along the distances typical of a reservoir's vertical extent is normally of the order of 0.1 million yr (assuming light oil viscosities — seeEngland et al., 1987), "significant" compositional differences could imply the presence of barriers to flow within the reservoir. The problem of what constitutes a "significant" difference will be discussed below. This is, however, a critical problem in appraisal applications because it sets the threshold at which barriers/baffles are invoked by the interpreter.

In the Forties Field (UK North Sea) it was possible to demonstrate that two parts of the same field have different GORs and subsurface densities co-existing at the same depth. It was also possible to estimate the rate at which density differences are eliminated (both analytically and numerically) and show that the two parts of the field must be separated by a barrier. This was confirmed by production history (England et al., 1995).

Stainforth (2004) proposed a different picture of reservoir filling. In his model, gravitational and diffusive effects are not considered efficient on a geological timescale. Vertical gradients are "gravitationally over-stable" – i.e. stronger than can be explained by the equilibrium effect of thermal and gravitational fields. The range of API seen in the trap reflects the range generated during source rock evolution — the different fluids find their own level in the reservoir without significant mixing. This interesting new model could be regarded as an end-member and will doubtless be the subject of future research involving detailed geochemical examination of reservoir fluids within vertical columns.

In most reservoirs, samples from different wells have been collected at different depths. If the magnitude of vertical gradients is not well understood, it is extremely difficult to interpret any (possible) lateral well-to-well gradients that may in reality reflect one field-wide vertical gradient sampled at different depths. This is one of the reasons why the understanding of vertical gradients is so important for reservoir appraisal and the application of reservoir geochemistry in particular.

Lateral Gradients and their Origin

England et al. (1987) proposed that lateral compositional gradients could be inherited from the way in which a field fills from its source rock kitchen(s). These differences are not totally eliminated due to mixing that is slow on a geological timescale (or of course the presence flow barriers). Some of these effects are summarised in Fig. 1.

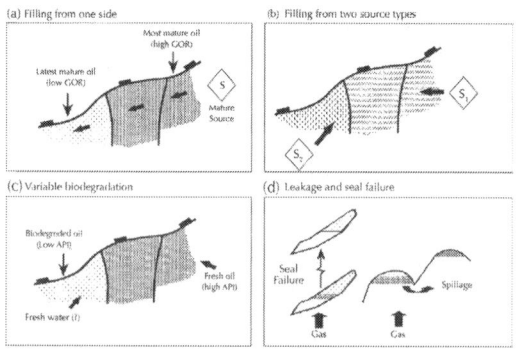

Figure 1: Various geological mechanisms that may induce compositional variation within petroleum accumulations.

In many cases filling occurs preferentially from one side Fig. 1a, or from two sources Fig. 1b. Present day samples taken from wells nearest to a mature source rock kitchen have increased maturity (evidenced from higher GOR, lower density and more mature biomarker signatures).

Another very common mechanism involves biodegradation of an oil field (Fig. 1c). Because of the sensitivity of bacteria to temperature and the presence of oxidants such as sulphate ions and formation waters, different parts of a field often experience different degrees of biodegradational attack. In the presence of geologically slow mixing or significant barriers, different samples can have greatly differing compositions, particularly with respect to API gravity, GOR and wax content (Larter and Aplin, 1994 and Smalley et al., 1997).

Finally (Fig. 1d) in basins where reservoirs are prone to seal failure or gas influx and subsequent spillage, significant vertical differences in petroleum composition may be induced due to phase fractionation and the effect of substantial ingress of gas, (Thompson, 2004).

It should be remembered that an apparently "lateral" gradient may be the expression of a true vertical gradient, due to the sampling of different wells at different depths. It is also possible that a combined lateral temperature gradient combined with the Earth's gravitational field can lead to compositions that vary both vertically and laterally even in the absence of flow barriers (Jacqmin, 1990).

Measurements on Cores

Measurements on cores can be based on petroleums and formation waters extracted in the laboratory. The advantage of core measurements is the much higher spatial resolution in which samples can be taken at centimetre intervals if necessary.

Fluid Inclusions

The study of fluid inclusions offers the possibility of examining petroleums and waters that were trapped in inclusions at some time in the geological past. This is invaluable in understanding the overall migration history of a field or set of fields, as it gives a unique opportunity to test models of petroleum generation, expulsion and migration.

Karlsen et al. (2004) investigated petroleum migration and overpressure evolution in the Haltenbanken, offshore Norway. By comparing the petroleum geochemical characterisation of current petroleums from reservoir core extracts with paleo-petroleums trapped in fluid inclusions, they demonstrated a very complex evolution of petroleum in the reservoirs. Both currently filled and currently dry structures in the Smørbukk field area were initially charged with black oil some 50–70 million yr ago. The currently dry structures were then filled with gas condensate prior to seal failure and leakage of the petroleum from the reservoir. The leakage events were interpreted as being caused by overpressure development related to diagenesis in the Smørbukk fault zone. In an appraisal situation this type of study could be very useful in elucidating the risks associated with undrilled segments (such as fault compartments) around an existing discovery.

Tar Mats

"Tar mats" are zones of heavy oil, rich in asphaltenes that are generally regarded as immobile on a production timescale. A prime reservoir engineering significance is the need to recognise certain reservoir intervals as containing petroleum saturated rock (e.g. from electric logs) that will nevertheless not produce oil under normal circumstances. Detailed mapping of tar mats in core can often reveal extreme vertical variability, with "mini-tar mats" – occurring at the bases of permeable intervals – that may reveal information about migration and potential reservoir baffles. More details are given in Wilhelms and Larter (1995) and in the discussion in Carpentier et al. (1998) of the distribution and origin of tar mats in a field in Abu Dhabi.

Residual Salt Analysis

Residual salt analysis uses carefully obtained aqueous extracts of small samples of dried cores sampled at regular intervals. This has the effect of re-dissolving salts that were precipitated during core storage. Strontium isotope analysis is carried out on the extracted waters and used to provide a vertical log of the strontium isotope ratio. In many basins the strontium isotope ratio of the aquifers evolves with geological time due to changes in temperature and pressure. Changes in strontium isotope ratio within water legs, coinciding with possible barriers or baffles such as shale layers, can be used to elucidate the presence of flow barriers. Applications in water legs are very useful to the reservoir engineer because efficient water injection is essential to many developments.

The interpretation of strontium isotope profiles in the oil leg is more complex, as the isotope signature is "frozen" once high oil saturation is achieved during the reservoir filling process. However, some studies have reported the application of this method, e.g. Mearns and McBride (1999). Residual salt analysis has the advantage that historical core samples can be used to make measurements on fields many years after the initial discovery.

Bulk Fluid Property Prediction

In multi-pay reservoirs, often encountered in deltaic settings, the multiplicity of thin reservoirs makes it economically hard to justify fluid sampling of all zones. In these circumstances, geochemical evaluation of side wall cores can offer a very valuable alternative method of evaluating API gravity and viscosity. The side wall cores do not generally produce enough fluid to make direct fluid property measurements — in any case they will be denuded of all gas and many of the lighter gasoline range components due to the depressurisation involved in taking samples to surface.

Depending on circumstance, the usual approach is to extract the petroleum from the core by either thermal desorption or solvent extraction followed by GC or GC-MS analysis (Jarvie et al., 2001). The analytical data can be used to examine reservoir continuity; API and viscosity can then be estimated by local or general correlations. One complicating factor is the uncertainty in how representative the material extracted from the core is of the truly mobile petroleum phase in the pore space. None the less, these methods can give very valuable information for the reservoir engineer. For example McCaffrey et al. (1996) used core extracts from the Cymric Field California to optimise the completion of new wells in the most favourable intervals. They calibrated the core extract information by cross-checking with bulk samples.

RESERVOIR ENGINEERING ISSUES DURING PRODUCTION

Once a field starts producing, the priority shifts to gathering an understanding of field performance as soon as possible. Key issues are the connectivity of producing and injecting wells and the overall quantity of oil that has been contacted by the producing wells. For example if the wells are draining a smaller than expected volume, then the production rate will decline at a rate faster than forecast. This may indicate errors in the assumptions used to calculate the oil initially in place (OIIP) or a lack of communication due to horizontal or vertical compartmentalisation. Similar problems occur if the gas or water injection wells are not in pressure communication with the producing wells.

Reservoir geochemistry can be of assistance by monitoring the chemical composition of oils from newly drilled producing wells on a regular basis from start-up onwards. Due to its relatively low cost, GC fingerprinting is often the technique of choice when using differences in oil and gas chemistry to assess the likelihood of compartmentalisation. This technique can also be applied to injection wells if they are drilled into a petroleum bearing zone.

Of course, there is no reason why more complex GCMS based techniques should not be used as well, although; in the majority of cases simple GC fingerprinting is all that is required. Depending on the problem being addressed, it is strongly recommended that adequate production samples are archived for later analysis.

In multi-zone reservoirs it is sometimes more economical to produce two or more zones through one production tubing "string". For proper reservoir management it is important to ascertain the relative volumes extracted from each reservoir at regular time intervals — this process is known as production allocation. This information is used to carry out material balance and history matching as well as to identify mechanical or other production problems.

Standard industry practice does not use chemical information for production allocation in multizone wells. However, the geochemical approach has the advantage of being very low cost compared with the standard procedure based on mechanical production logging tools. The cost of carrying out production logging runs to establish the relative production from different zones is high, especially for subsea wells where access to the well-head is expensive. This is therefore a potentially valuable application of reservoir geochemical techniques.

Geochemical production allocation relies on establishing a significant difference between the oils associated with the zones in question. A totally empirical evaluation of whole oil chromatograms can be very effective in the gasoline range, and is referred to as "GC fingerprinting". Isotopic composition or more complex GC-MS parameters can be equally effective. Nicolle et al. (1997) have demonstrated production allocations in the Thamama reservoirs, Abu Dhabi. They used peak ratios from the GC fingerprints and demonstrated agreement within 6% when compared to conventional measurements. Kaufman et al. (1990) gave other examples from the Gulf of Mexico and demonstrated an application that identified corrosion problems in production strings that had caused production from one zone to cease. Peters and Fowler (2002) also give more details on reservoir management applications.

INTERPRETATIONAL AND MEASUREMENT ISSUES

In this section I somewhat arbitrarily divide reservoir geochemical techniques into those that are "reasonably well established" and those that are "less well established", where further developments are needed to allow them to be widely and routinely applied.

Reasonably Well Established

Use of Chemical Differences for Appraisal and Production Allocation

In some geological basins, sand-to-sand and well-to-well lateral compositional differences are sufficiently large that they can readily be used to address issues such as the presence or absence of reservoir compartmentalisation. (This situation often applies in Tertiary deltas such as the Niger Delta, where very variable petroleum compositions are found vertically and laterally.) It is important to acknowledge that these techniques have been applied for long enough to establish a track record of success and failure of predicted and actual outcomes compared against production data. Applications are widespread, including both appraisal and production allocation (e.g. Kaufman et al., 1990, McCaffrey et al., 1996 and Peters and Fowler, 2002). As in any application to reservoir continuity, there is still the question of where to set the "threshold" between compositional differences that imply barriers and those that do not. In the absence of firm theoretical guidelines, these areas have sufficient field data to use as a basis for setting thresholds, comparing with production data and assessing the risk of inaccurate prediction.

One important factor in the successful applications is the complexity of the deltaic petroleum systems involving multiple reservoirs, often involving faulting, biodegradation and gas flushing and leakage. This can lead to a highly variable petroleum "feeding" into the reservoirs, which makes it much easier to apply techniques such as fingerprinting because there is a large "signal" to deal with. The extreme variability of the petroleums also implies that two samples with the same composition have a reasonably high probability of being in communication, though identical compositions do not necessarily demonstrate connectivity — it could just be coincidental (England, 1990). (Additional non-chemical data should of course also be used to help any interpretation).

Fully Integrated Studies

In many cases, successful integrated studies have been carried out where a broad range of data is interpreted in the geological context of the field concerned. For example GC, GC-MS, strontium isotope, core extract, formation pressure and PVT data are all combined in order to build up a consistent picture of the fluids and the drilled (and undrilled) reservoirs containing them. As many authors have stressed, an integrated study, combining all available data is more likely to be successful than is a narrow, purely geochemical approaches (e.g. Smalley and Hale, 1996).

In favourable cases, well test or formation pressure measurements allow the more subtle geochemical differences to be calibrated, so that the geochemical differences can be related to the reservoir "plumbing". In any event, even where the interpretation during appraisal is difficult, once some production data have been obtained from the field and potential barriers identified, this information can be fed back into the interpretation of fluid differences.

Core Based Measurements

The use of strontium isotopes from core extracts has been validated in several cases — however, the complications of understanding basinal water chemistry evolution and the subtle effects of reservoir filling mean that it is not ideally suited for application unless it is part of a well-integrated study.

All core based measurements suffer from the complication of drilling mud which is inevitably introduced into the samples. Although the interference can be reduced, particularly if mud samples are available, interpretations based on core extracts will always be less certain than those based on bulk fluid samples. (The problem of mud contamination is also a factor when interpreting any type of bulk fluid samples where sampling methods have not totally eliminated contamination — especially for wireline samples).

Less Well Established

Ab Initio Prediction of Thermo-gravitational Gradients

The difficulty of predicting the equilibrium vertical compositional gradient in petroleum reservoirs significantly hinders our ability to interpret compositional difference during appraisal. Not only do we lack a complete theoretical understanding of the rates and magnitudes of these processes, but there is only a limited empirical understanding of the range of gradients seen in natural systems. We also lack a definitive means to distinguish thermo-gravitational from non-thermo-gravitational gradients (e.g. by detailed chemical measurement).

Understanding the Controls on Compositional Heterogeneity during Reservoir Filling from the Petroleum System

Work is only recently beginning on including the effects of reservoirs filling and mixing into basin models. This makes it difficult to predict under which circumstance we would expect a highly variable or a homogenous hydrocarbon charge or vice-versa. Stainforth (2004) is one of the first authors to make this connection explicitly, though there is evidence from fluid inclusions of the wide range of compositions passing through the Smøbukk area, (Karlsen et al., 2004). Linking petroleum filling to structural evolution is very important. For example a field might fill and thoroughly mix over geological time resulting in a laterally homogenous petroleum composition. Late-life faulting would leave the petroleum compositions unchanged (Fig. 2) and thus any fingerprint-based interpretation suggesting that the field was un-compartmentalised would probably be erroneous. This hypothetical example shows the importance of interpreting fluids data in conjunction with the relevant physical and geological factors.

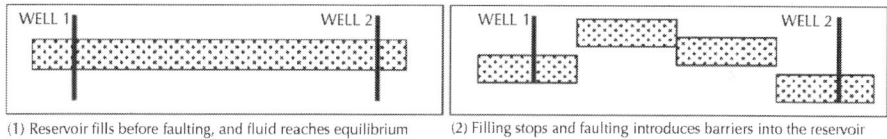

(1) Reservoir fills before faulting, and fluid reaches equilibrium   (2) Filling stops and faulting introduces barriers into the reservoir (no fingerprint differences detectable)

Figure 2: Representation of a situation where identical compositions measured in wells 1 and 2 do not necessarily imply that there are no current day barriers to flow. Note: This assumes that small scale chemical differences among the fluids once compartmentalised have not developed over geological time.

Despite the existence of simple order-of-magnitude estimates of the rates of reservoir equilibration processes for compositional and pressure differences, it is rare for these considerations to be part of interpretations of compositional differences. For example, mixing rates differ by orders of magnitude between heavy oil reservoirs and gas condensate accumulations (Smalley et al., 2004).

CONCLUSIONS

The "reasonably well established" applications described above in Section 4.1 all depend on reliable geochemical analyses supported by extensive field testing. In successful applications, geochemical observations are fully integrated with all available PVT data, pressure measurements, and geological, geophysical and other subsurface information to give an integrated picture of fluid bodies. However, because the theoretical underpinning of reservoir geochemistry is not completely developed, it is often difficult to extend the methods to new areas and new situations without extensive testing.

Because of our incomplete understanding of the processes involved, at present we do not have a straightforward way of selecting good analogues to assist in our interpretations. For example, should they be selected on the basis of source rock evolution, thermal gradient, degree of uplift, reservoir depositional environment, or any one of a large number of other possible criteria? In the absence of a full understanding, the selection of analogues is difficult — except from closely neighbouring reservoirs with similar properties and geological histories.

This reinforces the remarks of Westrich et al. (1999), mentioned in the introduction, that the major barrier to further applications of reservoir geochemistry in the area of reservoir engineering relates to the issue that "We need to know when it works, when it does not and why".

Some major issues that need to be addressed to broaden applications of reservoir geochemistry include:

- When interpreting any kind of geochemical or PVT data, the interpreter needs to assign a "threshold" above which a (reservoir engineering timescale) barrier is implied by the geochemical data. In other words, once it is certain that a compositional difference is above the analytical/sampling uncertainty, how definite is the interpretation of a barrier? Where the rates of mixing are not fully understood, a barrier could be suggested from compositional differences alone, where none exists in reality — a "false positive".

- In many ways the opposite of (1) is the problem of "false negatives". If two samples from different parts of a reservoir are chemically identical, is it sufficient evidence to strongly suggest connectivity? The similarity of composition could be accidental — for example two fault blocks may have received very similar migrating petroleum from a very homogenous source rock kitchen. Alternatively, a previously well-connected reservoir (at time of fill) may have undergone later faulting (without any subsequent chemical change occurring).

- Another significant impediment to the wider acceptance of reservoir geochemical methods is the variety of chemical measurements, parameters and ratios that are used in the interpretation of fluid differences. For example one or more type of data may be used (e.g. GC-MS, GC fingerprints, isotopes and PVT data). A variety of methods from multidimensional statistical analysis, simple cross-plots, GC-fingerprint radar plots or simple visual inspection of GC traces may be used to present the evidence of compositional difference. Since there is a lack of consistency in the type of data or interpretation used for predicting reservoir compartmentalisation it is not surprising that non-specialists may occasionally view the process as somewhat mysterious.

- A lack of published case histories is also a difficulty. Ideally, case histories should contain engineering data to support or disprove the occurrence of the flow baffles/barriers predicted by reservoir geochemical predictions. This is not possible for fields that are not yet producing oil or gas. For these fields, case histories should ideally be provided at a later date. However, where the gap between discovery and production can be in excess of 3–5 yr, there may be practical problems in doing this. (In addition many workers are naturally reluctant to highlight imperfect predictions that may have been made several years earlier — although these results are essential in developing our interpretive capability).

- Many workers have suggested coupling basin models and reservoir models in order to better understand the processes that create and (subsequently reduce and eliminate) compositional differences. The challenges presented by this approach are many — however progress is to be expected over the next few years. Some of the processes that are currently not usually simultaneously modelled in commercial basin or reservoir models include (NB many research codes may model one or two of these processes):

   a.   Molecular diffusion (thermal and chemical)

   b.   Vertical segregation gradients under thermal and gravitational fields

   c.   The rate of establishment of the above gradients

   d.   Thermal convection

   e.   The rates of biodegradation

   f.   The rates of leakage through cap rocks

   g.   Reservoir filling (in reservoir models)

   h.   The evolution of 3D geometry including the effects of sealing/partially sealing faults. (in reservoir models)

The challenge in simultaneously modelling the processes involved is considerable. Doubtless much progress will be achieved if one or two of the most important processes can be identified and modelled in particular basin settings. Stainforth (2004) was able to make some interesting observations on a set of reservoirs by making comparisons of the ranges of GORs predicted to be expelled from particular sourcing systems. This suggests that progress may be made without the need for a completely coupled modelling approach.

Reservoir geochemistry has already established itself as an extremely useful tool in the reservoir engineering domain. Greater acceptance depends on an increasingly systematic approach to the methods used and the interpretations supplied. A wider collection and dissemination of case histories, together with the predictions made, tested against production data will allow more informed decisions to be made. Advances in understanding the details of reservoir filling and mixing should allow a better appreciation of which techniques are likely to work best, given the geological setting involved. This would allow us to understand what works where, and why.

ACKNOWLEDGMENTS

I acknowledge the support and encouragement of colleagues and for BP for permission to publish. I thank J. Curiale and H. Halpern for their careful and constructive reviews.

REFERENCES

1. Carpentier, B., Badr, A.E.R., Mueller, H.W., Al Aidarous, A.A., AlBaker, S., 1998. Distribution and origin of tar mats in the Thamama Zone B of an Abu Dhabi field, ADIPEC, 8th Abu Dhabi International Petroleum Exhibition and Conference (11–14 October 1998) Abu Dhabi, U.A.E. SPE, p. 49472.

2. Cubitt, J.M., England, W.A., 1995. The Geochemistry of Reservoirs. Geol. Soc. Lond. Spec. Publ. 86.

3. Cubitt, J.M., England, W.A., Larter, 2004. Understanding petroleum reservoirs: towards an integrated engineering and geochemical approach. Geol. Soc. Lond. Spec. Publ. 237.

4. Dake, L.P., 1985. Fundamentals of Reservoir Engineering (Developments in Petroleum Science). Elsevier ISBN: 044441830X.

5. England, W.A., 1990. The organic geochemistry of petroleum reservoirs. In: Durand, B., Behar, F. (Eds.), Advances in Organic Geochemistry 1989. Organic Geochemistry, vol. 16, pp. 415–425.

6. England, W.A., Mackenzie, A.S., Quigley, T.M., Mann, D.M., 1987. The movement and entrapment of petroleum in the subsurface. J. Geol. Soc. London 144, 233.

7. England, W.A., Muggeridge, A.H., Clifford, P.J., Tang, Z., 1995. Modelling density-driven mixing rates on a geological timescale with application to the detection of barriers in the Forties Field (UKCS). In Cubitt and England (1995) ibid.

8. Ghorayeb, Firoozabadi, 2000. Numerical study of natural convection and diffusion in fractured porous media. SEPJ 12 (March).

9. Hirchberg, A., 1988. The role of asphaltenes in compositional grading of a reservoir's fluid column. JPT 89 (Jan).

10. Høier, L., Whitson, C.H., 2001. Compositional grading — theory and practice. SPE Reserv. Eng. 525 (December).

11. Jacqmin, D., 1990. Interaction of natural convection and gravity segregation in oil/gas reservoirs. SPERE, 233 (May) Trans. AIME 289.

12. Jarvie, D.M., Morlelos, A., Han, Zhiwen, 2001. Detection of pay zones and pays quality, Gulf of Mexico: application of geochemical techniques. Gulf Coast Assoc. Geol. Soc. Trans. LI 151–160.

13. Karlsen, D.A., Skeie, J.E., Backer-Owe, J.E., Bjørlykke, K., Okstad, R., Berge, K., Cecchi, M., Vik, E., Schaefer, R.G., 2004. In: Cubitt, J.M., England, W.A., Larter (Eds.), Petroleum Migration, Faults and Overpressure. Part 2. Case History: The Haltenbanken Petroleum Province, Offshore Norway.

14. Kaufman, R.L., Ahmed, A.S., Elsinger, R.J., 1990. Gas Chromatography as a development and production tool for fingerprinting oils from individual reservoirs: applications in the Gulf of Mexico. In: Schumaker, D., Perkins, B.F. (Eds.), Proceedings of the 9th Annual Research Conference of the Society of Economic Paleontologists and Mineralogists, October 1, 1990: New Orleans, pp. 263–282.

15. Larter, S.R., Aplin, A.C., 1994. Production Applications of reservoir geochemistry: a current and log-term view. SPE 28375, 105–113.

16. Larter, S.R., Carpentier, B., Lafargue, E., Huc, A.-Y., 1995. Reservoir geochemistry at a reservoir appraisal and management tool. An evaluation. AAPG Bull. 79 (8), 1228.

17. McCain, W.D., 1990. The Properties of Petroleum Fluids. Publ. Pennwell. 0878143352.

18. McCaffrey, M.A., Legarre, H.A., Johnson, S.J., 1996. Using Biomarkers to improve heavy oil reservoir management: an example from the Cymric Field, Kern County, California. Am. Assoc. Pet. Geol. Bull. 80 (6), 898–913.

19. Mearns, E.W., McBride, J.J., 1999. Hydrocarbon filling history and reservoir continuity of oil fields evaluated using 87Sr/86Sr isotope ratio variations in formation water, with examples from the North Sea. Pet. Geosci. 5, 17–27.

20. Nicolle, G., Boibien, C., ten Haven, H.L., Tegelaar, E., Chavagnac, P., 1997. Geochemistry: a powerful tool for reservoir monitoring. SPE 37804, 395–401.

21. Peters, K.E., Fowler, M.G., 2002. Applications of petroleum geochemistry to exploration and reservoir management. Org. Geochem. 33, 5–36.

22. Ratulowski, J., Fuex, A.N., Westrich, J.T., Sieler, J.J., 2000. Theoretical and experimental investigation of isothermal compositional grading. SPE 63084.

23. Sage, B.H., Lacey, W.N., 1939. Gravitational concentration gradients in static columns of hydrocarbon fluids. Trans. AIME 132, 120.

24. Schulte, A.M., 1980. Compositional variations within a hydrocarbon column due to gravity. SPE 9235.

25. Smalley, P.C., Hale, N.A., 1996. Early identification of reservoir compartmentalisation by combining a range of conventional and novel data types. SPE Form. Eval. 163–169 (Sept.).

26. Smalley, P.C., Goodwin, N.S., Dillon, J.F., Bidinger, C.R., Drozd, R.J., 1997. New tools target oil-quality sweetspots in viscous oil accumulations. SPE Reserv. Eng. 157–161 (August).

27. Smalley, P., England, W.A., Muggeridge, A.H., Abaciouglu, Y., 2004. In: Cubitt, J.M., England, W.A., Larter, S.R. (Eds.), Rates of reservoir fluid mixing: implications for interpretation of fluid data.

28. Smith, P.J., Hendry, D.J., Crowther, A.R., 1993. The quantification and management of uncertainty in reserves. SPE 26056.

29. Stainforth, J.G., 2004. In: Cubitt, J.M., England, W.A., Larter (Eds.), new insights into reservoir filling and mixing processes.

30. Thompson, K.F.M., 2004. In: Cubitt, J.M., England, W.A., Larter (Eds.), Interpretation of charging phenomena based on reservoir fluid (PVT) data.

31. Westrich, J.T., Fuex, A.M., O'Neal, P.M., Halpern, H.I., 1999. Evaluating reservoir architecture in the northern Gulf of

Mexico with oil and gas chemistry. SPE Reserv. Evalu. Eng. 2, 514.

32.   Wilhelms, A., Larter, S.R., 1995. Overview of the geochemistry of some tar mats from the North Sea and USA: implications for tar mat origin. In: Cubitt, J.M., England, W.A. and Larter (2004), p. 87.

Efficient Simulation of Hydraulic Fractured Wells in Unconventional Reservoirs

D.Y. Dinga, Y.S. Wub, and L. Jeanninc

aIFP Energies nouvelles, 1-4, Avenue Bois-Préau, 92852 Rueil Malmaison, France
bColorado School of Mines, Golden, CO 80401, USA
cGDF SUEZ, 92930 Paris La Defense, France

ABSTRACT

Reservoir simulators remain essential tools for improved reservoir management in order to optimize hydraulic fracturing design in unconventional low permeability reservoirs. However, the commonly-used simulator requires too much computational efforts for the description of local phenomena in the vicinity of the hydraulic fractures, and is practically infeasible for field applications, due to large number of gridblocks involved in a full-field simulation with many multi-fractured complex wells. Therefore, coarse grid simulations have been widely used, but the techniques for the coarse grid simulation of fractured wells need to be improved.

In this paper, we present efficient numerical methods to handle both long-term well performance and transient behavior simulations for hydraulic fractured wells in unconventional reservoirs with a coarse grid. To simulate correctly the long-term behavior, transmissibilities around the fracture and the connection factor between a fractured block and the fracture are computed, based on a (pseudo)-steady-state near-fracture solution. This approach provides not only an accurate long-term well production calculation, but also a correct pressure distribution in the near-fracture region. In unconventional reservoirs, the transient effects cannot be ignored, due to the very low reservoir permeability and near-well/near-fracture physical processes coupled modeling technique is used, and its efficiency is demonstrated with a tight such as fracturing fluid induced formation damage. In order to handle the transient effect, the -gas reservoir.

INTRODUCTION

With the increasing demand of hydrocarbon facing current energy shortage, the unconventional resources, such as tight gas and shale gas plays, are becoming more and more important. Significant progress has been made in the past decade for economic development of unconventional petroleum resources. Economic production from unconventional tight and shale gas reservoirs depends upon artificial well stimulation like hydraulic fracturing. Now, reservoir modeling and simulation become increasingly important in order to optimize and improve reservoir management of a tight or shale field by designing fracture length and spacing, and well spacing or infill well location or trajectory, etc.

From a numerical modeling point of view, a single hydraulic

fracture can be handled for research purpose using a very refined grid for flow simulation. However, unconventional reservoir developments include many wells with multi-hydraulic fractures. These fractures are long with a length of several hundred meters and narrow with a width of only several millimeters. Besides, fracturing in unconventional reservoirs induces usually a complex fracture network by reactivating reservoir natural fractures (see, for example, Delorme et al., 2013 and Norbeck et al., 2014). A fine grid numerical model to simulate these field cases requires too much CPU time and it is generally impractical or even impossible to perform fine grid simulations in field applications. So, there is a need to model hydraulically fractured reservoirs with coarse gridblocks.

To simulate hydraulic fractures with a reservoir simulator, some authors suggested using an equivalent wellbore radius or a negative skin factor with a coarse grid (see, for example, Lefevre et al., 1993). But this approach does not produce "elliptical shape" pressure distribution around the fracture and can generate a negative well connection factor for the reservoir simulator. Elahmady and Wattenberger (2006) proposed to use pseudo-permeabilities to simulate flow perpendicular and along the fracture directions. Zhou and King (2011) used an upscaling method to handle fractured wells in heterogeneous media. Many papers can be found in the literature to discuss the simulation of fractured wells with coarse grids and to propose some practical solutions (see, for example, Sadrpanah et al., 2006, Timur, 2008, Abacioglu et al., 2009 and Burgoyne and Little, 2012).

The problem of well modeling has been discussed for a long time. Peaceman, 1978, Peaceman, 1983 and Peaceman, 1993 model has been widely used for reservoir simulations. But that model has its limitations and cannot be used for the fractured well simulation. To improve the well modeling, Ding et al. (1995), Ding (1996) and Ding and Jeannin (2001) proposed to modify the near-well numerical scheme, based on a steady-state or a pseudo-steady-state flow regime, to adapt the singular flow behavior in the near-well region with coarse grid simulations. This concept was also extended for the modeling of fractured wells by Ding and Chaput (1999).

For unconventional reservoir simulations, it is essential to model correctly long conductive fractures such as hydraulic fractures. Li and Lee (2008) proposed to use a transport index (fracture connection index) between a matrix block and the fracture to handle embedded discrete fractures. They assumed that the pressure varies linearly in the normal direction to each fracture and they computed an average distance between a fractured block and the fracture for the transport index calculation. The same formula was used by Hajibeygi et al. (2011) in hierarchical fracture model for conductive fracture modeling and by Moinfar et al. (2013) for the simulation of hydraulically fractured unconventional reservoirs. In unconventional reservoir simulations, even with the presence of complex fracture networks, it is common to simulate large hydraulic fractures explicitly using a discrete fracture approach and homogenize short/diffused fractures for a single or dual continuum (see, for example, Moinfar et al., 2013 and Wu et al., 2013). The flow transfer between a grid block and a hydraulic fracture as well as between grid blocks near a large fracture needs to be simulated with precisions. However, assuming linear pressure variation is not very accurate near the fracture extremities and in the zone where several fractures are intersected, especially on large gridblocks. In this paper, we propose to compute the fracture connection index (FCI) based on steady-state or pseudo steady-state pressure solution with integral representation. Our approach can handle complex cases like flow modeling near fracture extremities, and it can also be extended to compute flow transports between grid blocks near a fracture. Examples are presented to handle hydraulically fractured wells in unconventional reservoirs. Note that some formation damage issues can also be handled through the modification of connection factors (or skin) between a fractured block and the fracture.

Another issue related to the fractured well simulation in unconventional reservoirs is the modeling of transient flow behavior. In a conventional reservoir, the transient period is generally very short, and a pseudo-steady-state flow regime is quickly reached in the near-well/near-fracture region. The transient behavior can generally be neglected in conventional reservoir simulations, except for transient well testing. However, the transient behavior cannot be ignored in low permeability unconventional tight and shale gas reservoirs. Artus and Fructus (2012) used very fine meshes to simulate transient behavior in shale-gas reservoirs and they showed the necessity to improve transmissibility calculation even for very fine mesh connections around the fractures, especially near the fracture extremities.

Coarse gridblock is generally not adapted to the modeling of transient behavior (the size of the gridblock is too large compared to variation length of physical variables such as pressure, saturation) and the description of near-well/near-fracture physical processes. To simulate the transient behavior, Blanc et al. (1999) proposed a transient well index, based on an analytical solution, to handle well testing problem for vertical wells. Al-Mohannadi et al. (2007) and Aguilar et al. (2007) used time-dependent well index to simulate horizontal and multilateral wells. Archer (2010) also shows the necessity of transient well index for the coarse grid simulation. The transient effect becomes particularly important for fractured wells in the unconventional reservoir due to long fracture length and low reservoir permeability or long transient flow period (Medeiros et al., 2007 and Ibrahim, 2013).

Transient effects associated with the near-well/near-fracture physical process are generally more complex. For a fractured well in an unconventional reservoir, the typical physical process is the fracturing fluid induced formation damage and cleanup (Bennion et al., 2000, Friedel, 2004, Ding and Renard, 2005,Cheng, 2012, Agrawal and Sharma, 2013 and Bertoncello et al., 2014) or possible non-Darcy flow inside the fracture towards the well (Wu, 2002). All these studies require the use of very fine gridblocks around the fractures, and it is difficult to investigate well production behavior in a full-field context. Besides, a full-field reservoir simulator might not have all the options to simulate detailed near-well/near-fracture physical processes. In order to take into account the transient flow and near-well/near-fracture physics on one side and long-term production behavior on the other side, we propose to use the coupled modeling technique (Ding, 2010 and Ding et al., 2012). A coarse grid model is used for the long-term production simulation, and a detailed near-well/near-fracture model is used to simulate near-well/near-fracture physics. These two models are linked through a time-dependent numerical productivity index and/or fracture connection index.

In this paper, we will present first the well model, based on a steady-state (or pseudo-steady-state) flow regime, for the simulation of fractured wells, which is suitable for long-term well productivity calculation. Then the transient behavior related to coarse gridblock size and near-well/near-fracture physical processes is modeled through the technique of coupled modeling. Examples are presented and show that these two combined approaches are well suited for the simulation of fractured wells in unconventional reservoirs.

COARSE GRID SIMULATION OF FRACTURED WELLS

Problems of Previous Fractured Well Modeling

Peaceman, 1978, Peaceman, 1983 and Peaceman, 1993 well model has been widely used in flow simulators. To compute the wellbore pressure (with a constrained well flow rate) or the well flow rate (with an imposed bottom-hole pressure), Peaceman (1978) introduced a numerical productivity index (PI) on a wellblock W_0 to relate the wellbore pressure Pw_0, the well flow rate Q_0 to the simulated wellblock pressure P_0, by

$$Q_0 = PI(P_0 - P_{w0}), \tag{1}$$

with the numerical *PI* given by

$$PI = \frac{C}{\ln(r_0/r_w) + S},$$

(2)

where C is a positive constant, depending on the well length and the permeability on the wellblock, S is the skin factor, rw is the wellbore radius, and r_0 is the equivalent wellblock radius, which depends on the wellblock size and the well trajectory on the wellblock. For example, for a fully penetrating vertical well in a 2D square gridblock, the equivalent wellblock radius is given by

$r_0 = 0.2 \text{ x}$

(3)

with Δx the gridblock size.

It is not difficult to find that Peaceman's formula is not suitable for the fractured well modeling. One solution for the simulation of fractured wells is using the concept of an equivalent wellbore radius, which equals 1/2 of the fracture half length, or a negative skin factor. However, several problems arise with the equivalent wellbore radius or a negative skin factor: (1) the wellblock cannot be too small. If the wellblock size Δx is smaller than 2.5 times of the fracture half length, the numerical PI becomes a negative value. For example, for a fracture with a half length of 100 m, the wellblock size cannot be smaller than 250 m. In terms of skin, it cannot be too small so that $\ln(r_0/r_w) + S$ is negative. This constraint limits many applications. (2) The elliptic pressure distribution in the vicinity of a fractured well is modeled by a radial one. (3) The simulated result is not accurate before a pseudo steady-state regime is reached around the whole fracture. The transient period is as long as the fracture length increasing. (4) The above equivalent wellbore radius is based on the infinite conductivity assumption. The flow inside the fracture is not considered. (5) This model is not valid in a heterogeneous media.

Improved Well Modeling

To improve the well modeling, Ding et al. (1995), Ding (1996) and Ding and Jeannin (2001) proposed to ameliorate the numerical scheme for near-well flow calculations. An accurate near-well flow approximation improves automatically the numerical *PI* formulation. In general, if a near-well steady-state or pseudo-steady-state solution is known, we can determine a suitable numerical scheme by identifying correct flow calculations. Considering a simplified two-point flux approximation scheme:

$$F_{ij} = T_{ij}(P_j - P_i), \tag{4}$$

where *Pi* and *Pj* are the pressures on the blocks *i* and *j* respectively, F_{ij} is the flux term across the edge Γ_{ij} (Fig. 1), the transmissibility term T_{ij} can be determined with a (pseudo) steady-state flow solution by

$$T_{ij} = \frac{F_{ij}^s}{(P_j^s - P_i^s)}, \tag{5}$$

where P_i^s and P_j^s are the (pseudo) steady-state pressure on the blocks *i* and *j*   respectively, F_{ij}^s is the (pseudo) steady-state flux across the edge Γ_{ij}. The pressure variation in the vicinity of the well is very big due to the singular near-well flow behavior. On the wellblock, we can define the wellblock pressure as a volumetric average pressure or a pressure at a distance r_0 from the well. Once the wellblock pressure is well defined, we can naturally determine the numerical *PI* for Eq. (1) by

$$PI = \frac{Q_0^s}{(P_0^s - P_w^s)}, \tag{6}$$

where Q_0^s is the well flow rate on the wellblock W_0 under a (pseudo) steady-state regime, P_0^s and P_w^s are respectively the wellblock pressure and the wellbore pressure under the (pseudo) steady-state regime. So, with this approach, the near-well flow can be correctly handled and the numerical PI can also be easily determined.

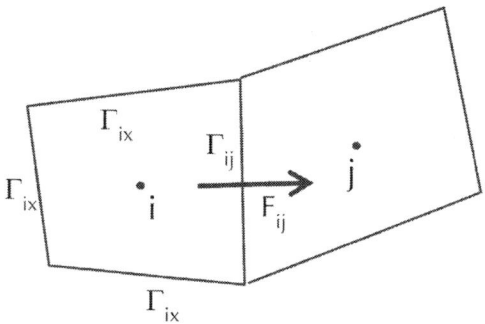

Figure 1: Flux approximation in the control-volume method.

Improved Fractured Well Modeling

For a fractured well, two problems should be considered in the flow modeling: flow from the reservoir to the fracture, and flow from the fracture to the well. The modeling of flow from the reservoir to the fracture is the most important one. Inspired by the well modeling with a numerical PI, we introduce a fracture connection index (FCI), which relates the pressure on a fracture block P_0, the fracture pressure P_{f0} on the same block and the flow rate entering to the fracture Q_{f0} by

$$Q_{f0} = FCI(P_0 - P_{f0}) \tag{7}$$

Considering a fully penetrated fracture (a 2D problem), a steady-state flow equation in the vicinity of the fracture in a homogeneous medium is given by

$$\frac{\partial^2 P^s}{\partial x_1^2} + \frac{\partial^2 P^s}{\partial x_2^2} = 0$$

(8)

with some boundary conditions at the fracture boundary Γ_f (Fig. 2), for example,

$$P^s(x) = P_f, \quad x = (x_1, x_2) \in \Gamma_f$$

(9)

for the Dirichlet condition, or

$$\frac{\partial P^s}{\partial n}(x) = Q_f, \quad x = (x_1, x_2) \in \Gamma_f$$

(10)

for the Neumann condition.

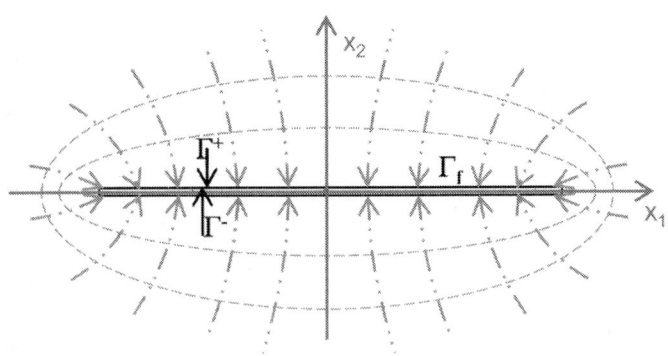

Figure 2: Modeling of steady-state flow towards the fracture.

presents a boundary integral method (sometimes also called semi-analytical method) to solve this steady-state problem. In particular, an analytical solution can be obtained with a constant potential condition (equivalent to the well-known uniform flux condition for the fractures). Once the steady-state solution is obtained, we can determine the transmissibility around the fracture by Eq. (5) and the fracture connection index (*FCI*) by

$$FCI = \frac{Q_{f0}^s}{(P_0^s - P_{f0}^s)}.$$

(11)

The pressure distribution, under a (pseudo) steady-state flow, depends mainly on the position of the fracture, and much less on the fracture conductivity. The transmissibilities and FCIs, obtained with a constant potential steady-state solution, can adequately be applied to the simulation of fractured wells with a wide range of conductivity values as shown in the examples hereafter.

It is generally recognized the importance of accurate modeling of flow transport between matrix and conductive fractures. Li and Lee (2008) assumed that the pressure varies linearly around the fracture and computed an average distance of a fractured block to the fracture for the matrix–fracture transfer calculation. The same approach is recently applied to flow simulation in unconventional reservoirs (Moinfar et al., 2013 and Norbeck et al., 2014) by considering the presence of main hydraulic fractures and a complex fracture network. However, using linear pressure assumption is not suitable for flow modeling near fracture extremities and in the zone of fracture intersections, especially for large gridblocks. Artus and Fructus (2012) illustrated the necessity of considering particular matrix–fracture flow near hydraulic fracture extremities. The formula proposed in this paper improves the matrix–fracture interaction for long conductive fractures, and examples below show the accuracy of our approach.

In heterogeneous media, no analytical or semi-analytical (pseudo) steady-state solution is available. However, we can use a technique similar to the near-well upscaling procedure (Ding, 1995 and Mascarenhas and Durlofsky, 2000) by solving a steady-state problem with a fine grid numerical simulation. The obtained fine grid solution is then used to compute the transmissibility around the fracture and fracture connection index (FCI) from Eqs. 5) and (11).

With this improved model for the fractured well modeling, the problems mentioned above are solved, in particular, (1) the concept of an equivalent wellbore radius is not used, so there is no limitation on the gridblock size. (2) The near-fracture pressure and flow distribution can be correctly handled, as the numerical scheme in the near-fracture region is modified to adapt the near-fracture flow behavior. (3) The transient behavior due to long fracture length can be partly reduced, because a pseudo-steady-state regime can be quickly attained in a scale between two neighboring gridblocks, much faster than in the whole fracture region. (4) This approach is not limited to an infinite conductivity fracture. Besides, this approach can be applied to a wide range of fracture conductivities, even if the flow regime is determined with a particular boundary condition (constant potential or other boundary conditions). (5) This concept can be extended to heterogeneous media.

Moreover, this approach can also be used for flow simulation with general discrete fracture network.

Flow inside the fracture has also to be considered, especially, for low conductivity fractures. In general, the simulation uses Darcy's law with a given fracture permeability to simulate flows inside the fracture. In the fracture plane, the well is considered as a sink/source at the intersection point between a horizontal well and the created fracture. So, the technique of well model (Peaceman, 1983 and Ding et al., 1995) can be directly applied to handle the radial flow inside the fracture towards a horizontal well. However, if a 1D grid is used inside the fracture (as shown in the examples hereafter), precaution should be taken, because a 1D grid cannot simulate correctly the radial flow behavior towards a horizontal well. Nevertheless, this has almost no impact on high conductivity fractured wells, as the pressure drop is very low along the fracture due to the high fracture conductivity. But errors can be significant in low conductivity fractures. One solution, as illustrated in the example below, is to introduce a skin factor or modify numerical PI for the connection between the wellblock (block containing the intersection of the well and the fracture inside the fracture plane) and the well to mimic a 2D radial flow in the fracture plane.

Remark

For the fractured well modeling, we use numerical PI to represent the connection between the well and the fracture, and *FCI* to represent the connection between the fracture and the reservoir.

Incorporation of Near-well/Near-fracture Physics and Coupled Modeling

Many physical processes may change near-well/near-fracture flow and formation properties. Concerning the fractured well, formation damage issue is greatly discussed (Bennion et al., 2000, Friedel, 2004, Ding and Renard, 2005, Agrawal and Sharma, 2013 and Bertoncello et al., 2014). In general, formation damage in fractured wells falls into two categories: damage inside the fracture and damage inside the formation. The damage inside the fracture can be caused by proppant crushing, proppant embedment, fracture plugging with chemicals and polymer residues. The damage inside the reservoir is classified by mechanical damage and hydraulic damage. The mechanical damage, which is generally characterized by a reduction of absolute permeability, includes polymer solids deposition near the fracture face, clay swelling, broken gel/fine migration, etc. The hydraulic damage occurs from the increase in water saturation caused by the fracturing fluid invasion into the porous media. It includes the hysteresis of relative permeability and high capillary pressure, water blocking, etc. The hydraulic damage reduces the gas flow in the zone, where the fractured fluids are invaded.

The formation damage can be characterized with skin factors. For a fractured well, if the formation damage is inside the reservoir, a skin is added to *FCI* term for the connection between the reservoir and the fracture. If the formation damage is inside the fracture, a skin is added to numerical *PI* term for the connection between the fracture and the intersected well. In Ding and Renard (2005), they presented a procedure for the determination of long-term numerical *PI*s for formation damage around a well. The similar procedure can be applied to determine *FCI*s for the formation damage near fracture faces. According to the physical problem, we will choose suitable indices (numerical *PI* and/or *FCI*) to integrate the damage effect.

Some physical processes alter progressively the near-fracture formation, and we cannot use constant skin factors (associated with numerical *PI* and/or *FCI*) to mimic this process. One solution is to use variable numerical *PI* and/or *FCI* through a coupled modeling (Ding, 2010). The coupled modeling is initially developed to update numerical *PI*s for a full-field coarse grid simulation. This technique is also suitable for updating *FCI*s for coarse grid fractured well simulations. The coupled modeling is particularly useful in the following two situations: (1) large gridblock size and low permeability; (2) near-well/near-fracture damage process acts in a long period.

In coupled modeling, data exchanges between the reservoir model and the near-fracture model are performed through updating numerical *PI*s and/or *FCI*s for the reservoir model and boundary conditions for the near-fracture model. The updated numerical *PI*s and/or *FCI*s correct inaccuracies in the coarse grid simulation. This technique can improve the modeling of transient effect considerably for fractured wells in full-field reservoir simulations.

NUMERICAL EXAMPLES

Example 1 – coarse Grid Simulation of a Single Fractured Well

In this example, we study how to handle a single fractured horizontal well with a coarse grid, and investigate the impact of different reservoir permeabilities, corresponding to a low permeability conventional reservoir, a tight gas reservoir and a shale gas reservoir, on the well performance simulation. Considering a half-geometry with respect to the fractured well as shown in Fig. 3. A fracture with half-length of 125 m and width 0.5 cm is first simulated with a very fine grid, starting with gridblock size of 0.25 cm around the fracture (Fig. 3a), and the reservoir thickness is discretized by 5 layers with gridblock sizes of 10 m. The horizontal well is drilled in the middle layer. This reservoir zone is also discretized with a uniform 2D coarse grid with block size of 50 m in the x and y directions and only 1 layer in the z-direction (Fig. 3b). For the coarse grid, the fractured well model presented above with a constant potential boundary condition is applied to handle the near-fracture flow by modifying the transmissibilities. It is assumed that only the fracture contributes the production. The reservoir water is immobile with irreducible water saturation of 0.2. Only gas flow is considered.

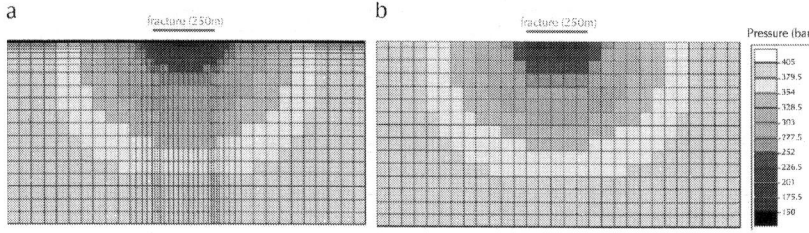

Figure 3: Discretization of a single fractured horizontal well (pressure distribution at 800 days with C_f=100 in a conventional low permeability reservoir): (a) fine grid system and (b) coarse grid system.

First, a low permeability "conventional" reservoir with permeability of 2 mD is considered. Fig. 4 presents a comparison between fine and coarse grid simulations on the gas production for a fracture permeability of 5 D, 50 D and 5000 D (or a dimensionless fracture conductivity C_f=0.1, 1 and 100). The dimensionless fracture conductivity is defined by

$$C_f = \frac{K_f\ w}{K_{res}\ X_f}.$$

(12)

with K_f the fracture permeability, K_{res} the reservoir permeability, w the fracture width, and X_f the fracture half length. The maximum gas production is limited to 10^6 m³/day. For the low fracture conductivity case, a significant difference between the coarse and fine grid simulations is observed, while for high fractures conductivities, the presented model is perfectly suitable for the coarse grid simulation. The transient period is less than 1 day due to high reservoir permeability.

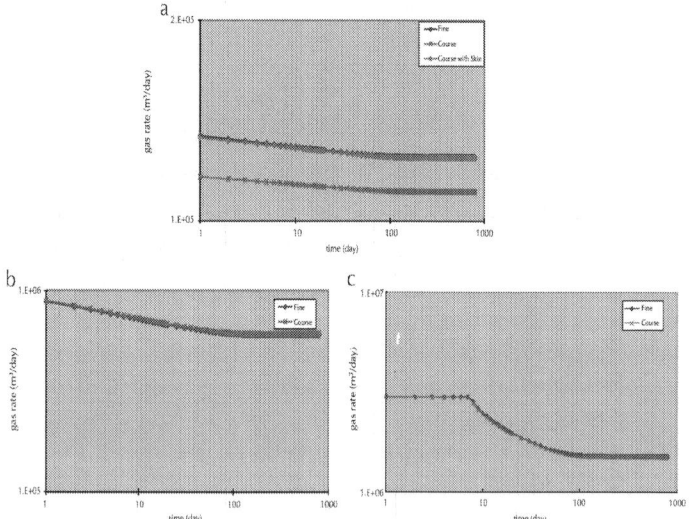

Figure 4: Comparison between coarse grid and fine grid simulations in a conventional low permeability reservoir: (a) C_f=0.1, (b) C_f=1 and (c) C_f=100.

In the coarse grid simulation, the fracture is discretized with a 1D grid. For high fracture conductivity, all gas in the fracture is quickly produced by the well, and the pressure drop near the well (inside the fracture) is small. So, the error on the well modeling between the fracture and the well is not observed. However, for a low conductivity case such as $C_f=0.1$, not all gas in fracture is immediately produced by the well, and the pressure drop inside the fracture is important. To correct errors in flow modeling from the fracture to the well, we adapt here a simple approach by introducing a skin factor for the connection between the fracture and the well (or modify the numerical *PI*). The skin (or modified numerical *PI*) is obtained using the inversion procedure (Ding and Renard, 2005) by comparing with a pseudo-steady-state solution from a fine grid simulation. The coarse grid simulation results with the inversed skin factor for $C_f=0.1$ (fracture permeability 5 D) are also shown in Fig. 4a. Very satisfactory result is obtained.

Fig. 3 presents also the pressure distribution at 800 days for $C_f=100$ with the fine and coarse grid simulations. These two pressure distributions are quite similar. Using the proposed fractured well model can correctly simulate fractured wells with a coarse grid system. The transient effect in a conventional reservoir is very small.

Now consider a tight-gas reservoir with a reservoir permeability of 0.02 mD. If we still use a fracture permeability of 5 D and 50 D, the dimensionless conductivity becomes 10 and 100 respectively, much higher than those in a conventional reservoir. Fig. 5 compares the gas production rate with the coarse and fine grid simulations. The coarse grid simulation provides generally accurate long-term production, but it remains inaccurate in early-time period due to large coarse gridblocks and lower reservoir permeability. The transient duration depends on the fracture permeability (or conductivity). In case of high fracture permeability, the transient period may last around 20 days, but the difference between coarse and fine grid simulations is not very large. Fig. 6 shows the pressure distribution at 800 days with fine and coarse grid simulations. The pressure distributions around the fracture are quite similar. Thus, we can simulate a fractured well in the tight-gas reservoir with a coarse grid with reasonable precisions, especially for long-term well production.

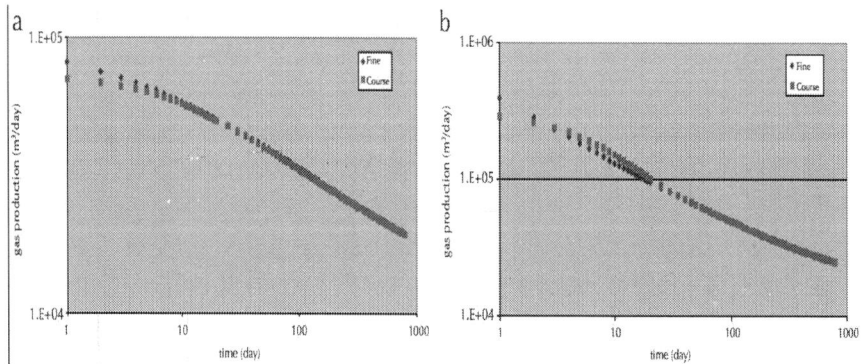

Figure 5: Comparison between coarse grid and fine grid simulations in a tight reservoir: (a) $C_f=1$ and (b) $C_f=10$.

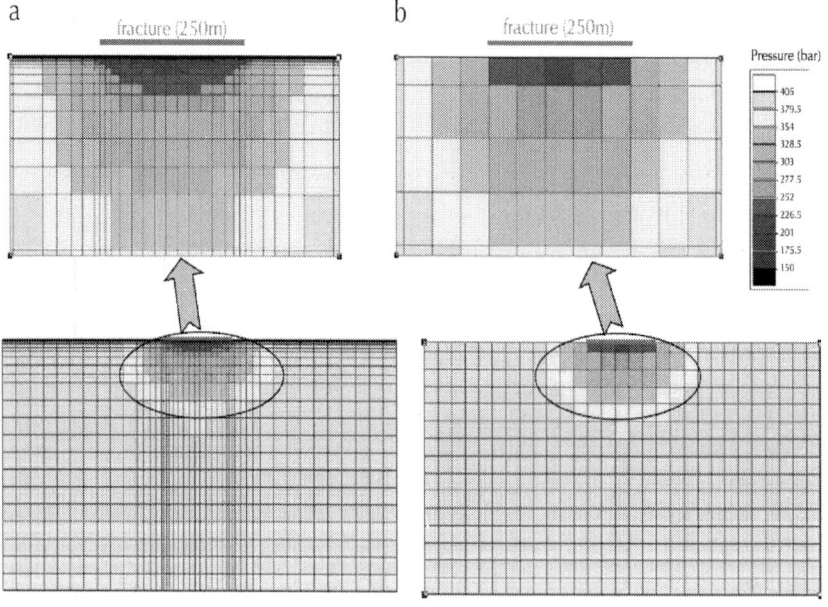

Figure 6: Pressure distribution with coarse grid and fine grid simulations in a tight reservoir: (a) fine grid simulation and (b) coarse grid simulation.

The transient effect seems limited in this example, because we consider only a single-phase Darcy flow. The transient effect can be much more important if other physical processes such as formation damage are considered. Fig. 7 shows a fine grid simulation by considering fracturing fluid induced formation damage for the case of K_f=5D. It is found that the errors related to the coarse grid simulation are more important, due to the presence of fracturing fluid, and the transient regime lasts longer. In Example 2 below, we will study details of the transient effect related to the fracturing induced formation damage and show how to correctly handle this effect with a coarse grid simulation using the couple modeling technique.

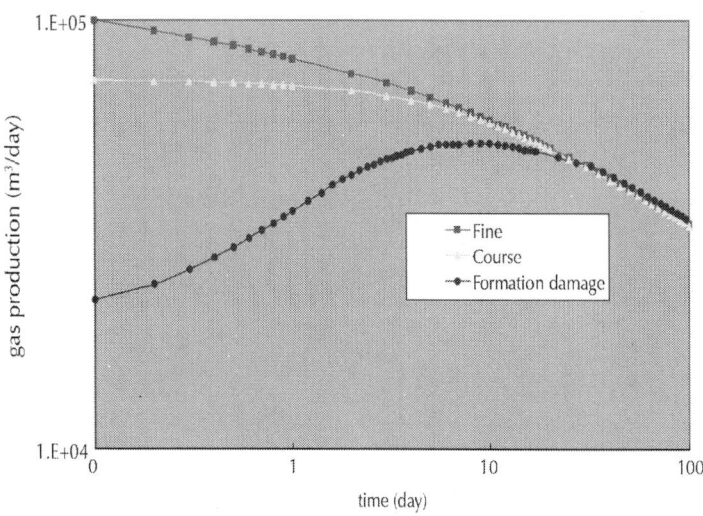

Figure 7: Transient effect considering fracturing fluid induced formation damage.

For the shale-gas or ultra-tight reservoir, the tight matrix permeability is extremely low. Let us consider a reservoir matrix permeability of 0.0001 mD. In case of a hydraulic fracture permeability of 5 D, the dimensionless conductivity is as high as 2000. Fig. 8a shows the simulation results with the coarse and fine grid systems. As the matrix permeability is ultra-low, a pseudo-steady-state regime cannot be reached before several years. At the beginning, gas production is underestimated with the coarse grid system due to its large gridblock sizes, which cannot simulate correctly the gas flow in the reservoir close to the fracture. This phenomenon lasts for a very long time. Later on, after the gas near the fracture is produced with the fine grid model, the coarse grid model starts to overestimate the gas production, because of the remaining gas in the near-fracture coarse blocks. Fig. 8b and c shows the simulation results with a lower fracture conductivity (C_f=160 and 16 respectively). For all simulations, the durations of transient period with the coarse grid system are almost the same. Only long-term production is slightly improved with low fracture conductivity.

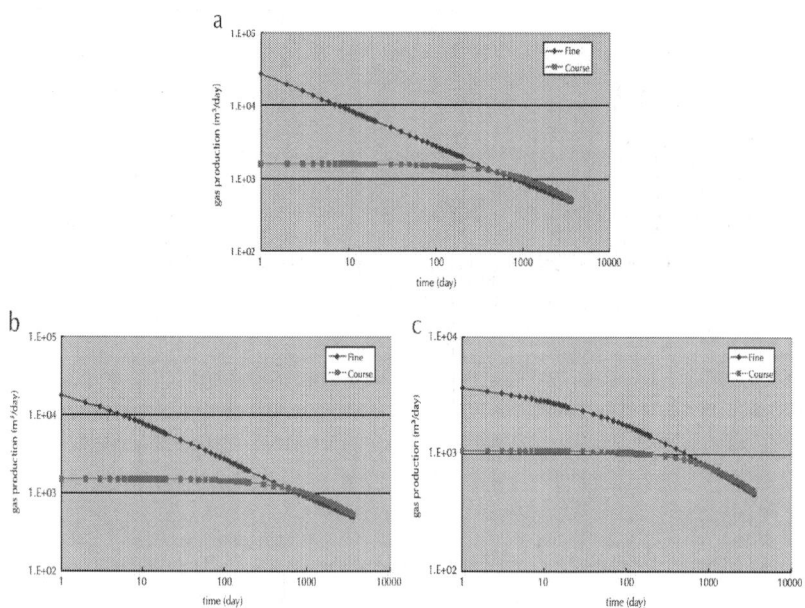

Figure 8: Comparison between coarse grid and fine grid simulations in a shale gas reservoir: (a) C_f=2000, (b) C_f=160, and (c) C_f=16.

The transient period is too long in the above simulations, and it is not suggested to use a transient numerical *PI* or *FCI* during a very long time period. A slight reduction of fracture block size can significantly decrease the transient time. Fig. 9 presents the simulation with one level grid refinement by subdividing a fracture block into two parts: a (fracture) block with a width of 10 m and a block with a width of 40 m. This subdivision reduces the transient period by almost a factor of 10. The coupled modeling can, therefore, be conveniently applied to simulate the transient effect if necessary.

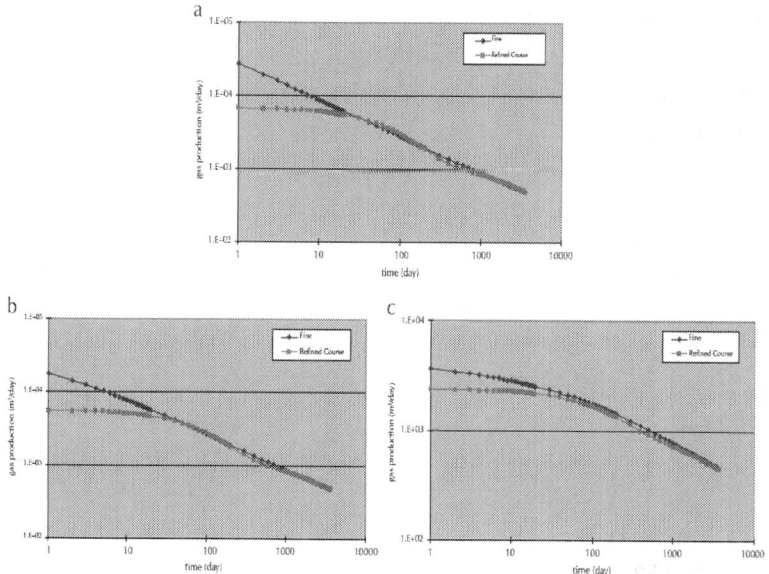

Figure 9: A slight refinement of fracture blocks in coarse grid simulations in the shale gas reservoir: (a) C_f=2000, (b) C_f=160, and (c) C_f=16.

The production with a fractured well generally follows two steps: flow from matrix to the fracture and then from the fracture to the well. In a shale-gas reservoir, the production is usually limited by the matrix–fracture transfer for the flow from the reservoir to the fracture. Transient effect with a coarse grid is extremely long. It has to be mentioned that the fracture system in a shale gas reservoir is much more complex than a single hydraulic fracture. Co-existence of fracture networks and natural fractures is usually needed to be considered. To simulate fractured wells in a shale-gas reservoir with a coarse grid, it is necessary to use a dual-porosity model and implement the fractured well modeling technique for dual-porosity simulations, which is our ongoing work.

In summary, the modeling technique presented in this paper can improve the simulation efficiency of fractured wells with a coarse grid system. When using a 1D grid in the fracture plane, a skin can be used to correct errors for the flow simulation from the fracture to the well. But this is not necessary if the fracture conductivity is high. For a low permeability "conventional" reservoir, transient effect can generally be neglected. For the tight gas reservoir, handling of transient period with a coarse grid simulation is an issue, especially when considering some damage processes. For the shale gas reservoir, the transient period depends greatly on the fracture block size, and the transient time might be extremely long. We should reasonably reduce the gridblock sizes for the fracture blocks in shale-gas reservoir simulations.

Example 2 – coarse Grid Simulation with Coupled Modeling

Considering a tight-gas reservoir of 1000 m×1000 m×50 m. The reservoir is water under-saturated: the initial reservoir water saturation of 0.2 is smaller than the irreducible water saturation of 0.28, which will induce water blocking in the fracturing fluid invaded zone. The relative permeability and capillary pressure curves are shown in Fig. 10. The average horizontal permeability is 0.02 mD and the vertical and horizontal permeability ratio is 0.5.

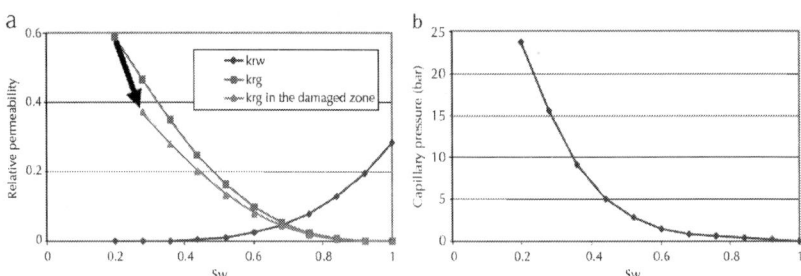

Figure 10: Relative permeability and capillary pressure: (a) hysteresis of gas relative permeability and (b) capillary pressure.

Two wells are drilled in this reservoir: one multi-fractured horizontal well with 3 transverse fractures (W1) and one single-fractured horizontal well (W2). Fig. 11a shows the coarse grid model and the location of these two fractured wells. All fracture half lengths are 50 m and its width is 1 cm. The dimensionless fracture conductivity in the multi-fractured horizontal well (W1) is 2, and it is 20 in the single-fractured horizontal well (W2). In all the wells, only fractures contribute directly to the gas productions. Due to water blocking, the gas relative permeability is reduced from 0.59 to 0.37 in the damaged zone (fracturing fluid invaded area). At the beginning of production, gas is produced together with the injected fracturing fluid, which reduces the gas flow rate in the cleanup period.

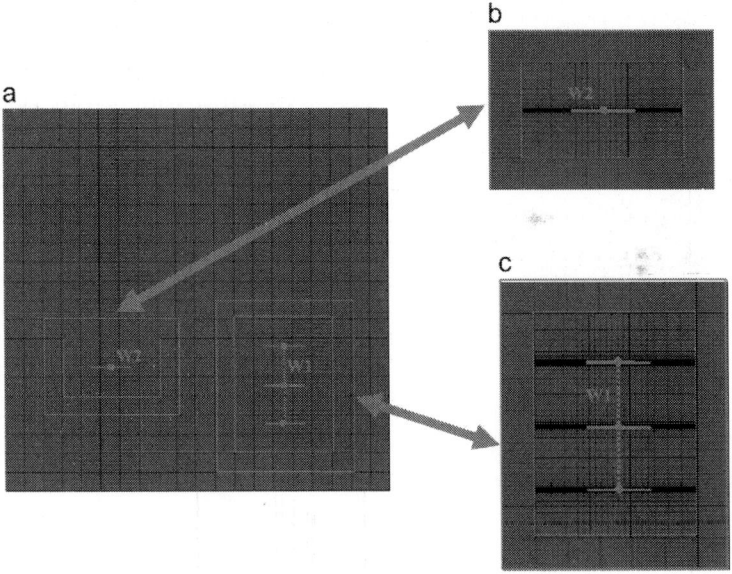

Figure 11: Coupled modeling of a two-well system: (a) full-field coarse grid, (b) near-fracture model for well W2, and (c) near-fracture model for well W1.

For the coarse grid simulation, the reservoir is discretized by 20 gridblocks in the x-direction, 20 gridblocks in the y-direction and

5 blocks in the z-direction with gridblock size of 50 m×50 m×10 m as shown in Fig. 11a. Transmissibilities around the fractures are modified by using Eq. (5) and the fracture connection indices FCI are given by Eq. (11) based on a constant potential steady-state flow.

Well W1 is drilled and fractured first. Well W2 is drilled and fractured one month later. A volume of 300 m³ of water-based polymer is injected to create each fracture, and the wells are shut in for two days before putting into productions. To simulate the fracturing fluid invasion and its cleanup, very fine gridblocks (sizes of several centimeters) are required around the fractures to get the reference solution, as the invasion depth is usually of the order of several to several tens centimeters. Fig. 12 presents a comparison of gas production between the coarse grid simulation with constant PIs and the reference solution. It is found that the coarse grid model cannot simulate the fracturing fluid invasion and the cleanup, due to large sizes of the fracture blocks. In fact, the initial reservoir saturation of 0.2 is lower than the irreducible water saturation of 0.28. Because of the large block size, all the injected fracturing fluid is blocked in the coarse gridblock and cannot be reproduced. A big difference is observed in early-time, due to the cleanup process and large gridblock sizes. The transient period lasts for around 200 days for well W1. Nevertheless, the long-term simulation is quite satisfactorily simulated with the coarse grid. Although some drilling fluid is still trapped inside the reservoir, it has not much impact on long-term well productions, because the flow is linear close to the fracture face and the damage in a small zone near the fracture does not have much impact on the linear flow (perpendicular to the fracture). This example also shows that the fracturing fluid induced formation damage has little impact on long-term gas production, but the cleanup period may last several months. It may also be much longer according to the reservoirs.

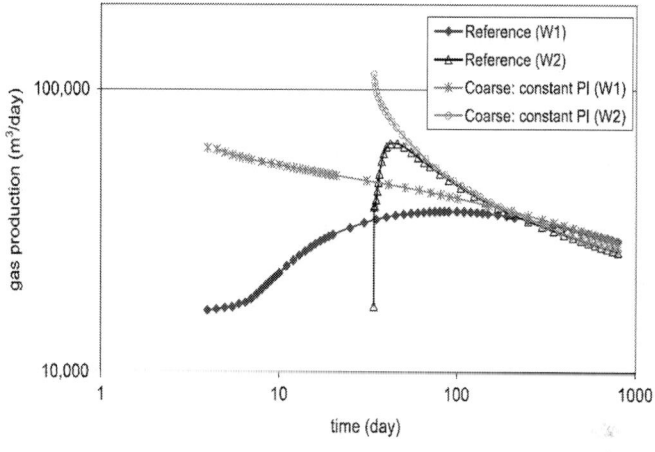

Figure 12: Coarse grid simulation with constant numerical PI.

To take into account the transient effects, the coupled modeling is used in the early time simulation when the well is fractured and starts the production. Fig. 11 also shows the fine grid models and their corresponding zones in the coarse grid. The coupled modeling for these two wells does not start at the same time, and the coupling periods are also different. Fig. 13 presents schematically the periods for the coupled modeling. The coarse grid full-field simulation is performed for a total of 800 days. The coupled modeling for well W1 starts with the injection of fracturing fluid at the 1st day. The duration of the coupling modeling is 200 days for this low conductivity multi-fractured well, because the cleanup time is long for a low conductivity fracture. In the same time, well W2 is fractured at the 31st day. So the coupled modeling for well W2 is started at the 31st day, and ended at the 61st day for a period of 30 days for this high conductivity fractured well. Table 1 presents the time steps for the data exchanges in the coupled modeling of these two wells. In this coupled modeling, we update the numerical *PI*s for the connections between the fractures and the well. Fig. 14 presents the simulation results of the coupled modeling. These results are very close to the reference solutions, which are obtained using local grid refinement with very fine gridblocks around the fractures.

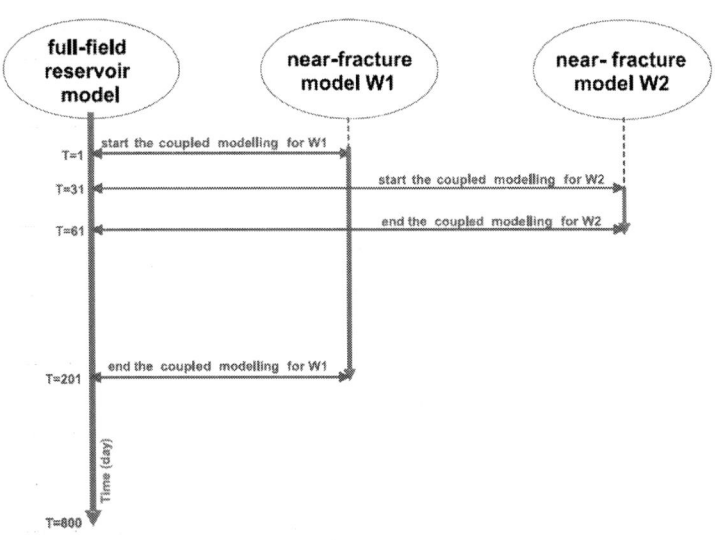

Figure 13: Time period for the coupled modeling.

Table 1: Time steps for data updating in the coupled modeling

Coarse grid full-field model	Time steps for the data exchange	
Time	Well W1	Well W2
(day)	(day)	
0–1	No coupling	No coupling
1–21	1	No coupling
21–31	5	No coupling
31–51	5	1
51–61	10	1
61–201	10	No coupling
201–800	No coupling	No coupling

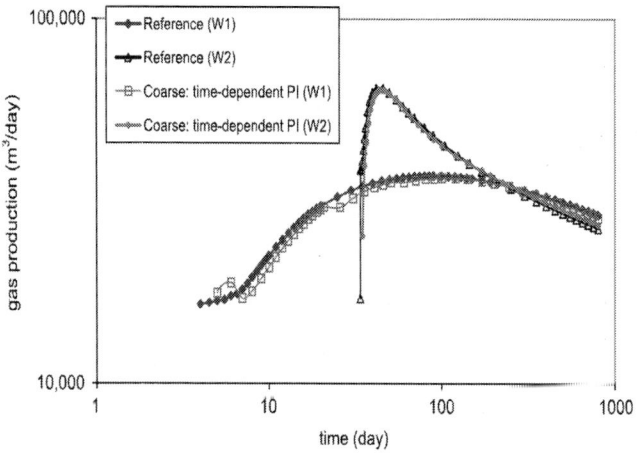

Figure 14: Coarse grid simulation with time-dependent PI.

This example shows that the proposed model, obtained by modifying the near-fracture transmissibility and the fracture connection factor *FCI* to adapt near-fracture flow behavior, gives very satisfactory solution for long-term well production. However, the differences between using the constant numerical *PI* and transient*PI* (or the reference model) are significant during the early time. To take into account near-fracture damage process, it is essential to use a coupled modeling, where the transient affects, related to the fracturing fluid induced formation damage and gridblock sizes, can be correctly handled through time-dependent *PI*s. The transient effect is small for high conductivity fractures, while it is large for low conductivity fractures. This example shows also the importance of a high fracture conductivity, since the single fractured well W2 with a conductivity of 20 can produce almost the same volume as the 3 fractured well W1 with a conductivity of 2.

DISCUSSIONS AND CONCLUSIONS

This paper studies the techniques for fractured well simulations in unconventional reservoirs with a coarse grid. A fractured well model, based on near-fracture transmissibilities and fracture connection indices determined from a steady-state flow regime, is proposed. This model is well suited for the simulation of long-term well production. If the near-well/near-fracture formation damage processes can be neglected, this model is suitable for the simulation of fractured wells in low permeability "conventional" reservoirs as well as unconventional tight-gas reservoirs. For shale-gas reservoirs, the transient period may be very long with a large coarse grid. A reasonable block size for fracture blocks (of the order of several meters) is required.

In homogeneous media, an analytical steady-state solution can be obtained using constant potential condition (equivalent to the uniform flux condition), and the fractured well modeling based on that solution is accurate enough for a wide range of fracture conductivity values. In heterogeneous media, the upscaling procedure, based on a fine grid simulation, can be used. Satisfactory simulation results make us confident to generalize this approach to more complex discrete fracture networks.

Accurate simulation of flow inside the fracture plane is also required for low conductivity fractures, especially if a fracture is simulated with a 1D grid. One practical solution is using a skin factor or modifying the numerical *PI* with an inversion procedure.

Transient flow may occur during early-time due to large gridblock, low permeability and some near-well/near-fracture physical processes. The transient period may be very long and important according to the reservoir and damage process. To handle the transient flow simulation with a coarse grid system, time-dependent numerical *PI*s and/or *FCI*s are necessary. The coupled modeling is a useful technique to provide time-dependent numerical *PI* for coarse grid simulations, even if the coarse grid model does not dispose the option to simulate the near-well/near-fracture physical processes.

REFERENCES

1. Abacioglu, Y., Sebastian, H.M., Oluwa, J.B., 2009. Advancing reservoir simulation capabilities for tight gas reservoirs. In: SPE 122793 Presented at the 2009 SPE Rocky Mountain Petroleum Conference, 14–16 April. Denver, CO, USA.

2. Agrawal, S., Sharma, M., 2013. Liquid loading within hydraulic fractures and its impact on unconventional reservoir productivity. In: SPE 168781/URTeC 1580636, Unconventional Resources Technology Conference held in Denver, Colorado, USA, 12–14 August.

3. Aguilar, C., Ozkan, E., Kazemi, H., Al-Kobaisi, M., Ramirez, B., 2007. Transient behavior of multilateral wells in numerical models: a hybrid analytical– numerical approach. In: SPE 104581 Presented at the 15th SPE Middle East Oil & Gas Show and Conference, 11–14 March. Bahrain.

4. Al-Mohannadi, N., Ozkan, E., Kazemi, H., 2007. Grid-system requirements in numerical modeling of pressure-transient tests in horizontal wells. In: SPE Reservoir Evaluation & Engineering, April, pp. 122–131.

5. Archer, R.A., 2010. Transient well indices: a link between analytical solution accuracy and coarse grid efficiency. In: SPE 134832 Presented at the SPE ATC&E, Florence, Italy, 19–22 Sept.

6. Artus, V., Fructus, D., 2012. Transmissibility corrections and grid control for shale gas numerical production forecasts. Oil Gas Sci. Technol. 67 (5), 805–821.

7. Blanc, G., Ding, D.Y., Ene, I.-A., Estebenet, T., 1999. Transient productivity index for numerical well test simulations. In: Schatzinger, R, Jordan, J.F. (Eds.), Reservoir Characterization, vol. 71. AAPG Memoir, pp. 163–174.

8. Bennion, D.B., Thomas, F.B., Ma, T., 2000. Recent advances in laboratory test protocol to evaluate optimum drilling, completion and stimulation practices for low permeability gas reservoirs. In: Paper SPE 60324, Presented at the SPE Rocky

Mountain Regional/Low-Permeability Reservoirs Symposium and Exhibition, Denver, CO, March 12–15.

9.   Bertoncello, A., Wallace, J., Blytin, C., Kabir, C.S., 2014. Imbibition and water blockage in unconventional reservoirs: well management implications during flowback and early production. In: SPE 167698, Presented at the European Unconventional Resources Conference, Austria, 25–27 February.

10. Burgoyne, M.W., Little, A.L., 2012. From high perm oil to tight gas—a practical approach to model hydraulically fractured well performance in coarse grid reservoir simulators. In: SPE 156610 Presented at the SPE Asia Pacific Oil and Gas Conference & Exhibition, Perth, Australia, 22–24 Oct.

11. Cheng, Y., 2012. Impact of water dynamics in fractures on the performance of hydraulically fractured wells in gas-shale reservoirs. Can. Petrol. Technol., 143–151.

12. Delorme, M., Daniel, J.M., Kada-Kloucha, C., Khvoenkova, N., Schueller, S., Souque, C., 2013. An efficient model to simulate reservoir stimulation and induced microseismic events on 3D discrete fracture network for unconventional reservoirs. In: Paper SPE 168726 Presented at the Unconventional Resources Technology Conference, Denver, CO, USA, 12–14 August.

13. Ding, Y., Renard, G., Weill, L., 1995. Representation of wells in numerical reservoir simulation. In: SPE 29123 Presented at the 13th Reservoir Simulation Symposium, San Antonio, USA, 12–15 Feb.

14. Ding, Y., 1995. Scaling-up in the vicinity of wells in a heterogeneous field. In: Paper SPE 29137 Presented at the 13th Symposium on Reservoir Simulation, San Antonio, USA, 12–15, Feb.

15. Ding, Y., 1996. A generalized 3D well model for reservoir simulation. SPE J. 1 (4), 437–450. Ding, Y., Chaput, E., 1999. Simulation of hydraulically fractured wells. Petrol. Geosci. 5, 265–270.

16. Ding, D.Y., Jeannin, L., 2001. A new methodology for singularity modeling in flow simulations in reservoir engineering. Comp. Geosci., 93–119.

17. Ding, D.Y., Renard, G., 2005. Evaluation of horizontal well performance after drilling\ induced formation damage. J. Energy Resour. Technol. 127, 257–263.

18. Ding, D.Y., 2010. Modeling formation damage for flow simulations in reservoir scale. SPE J. 15 (3), 737–750.

19. Ding, D.Y., Langouet, H., Jeannin, L., 2012. Simulation of fracturing induced formation damage and gas production from fractured wells in tight gas reservoirs. In: SPE 153255 Presented at the 2012 SPE East Unconventional Gas Conference and Exhibition, Abu Dhabi, 23–25 Jan.

20. Elahmady, M., Wattenberger, R.A., 2006. Coarse scale simulation in tight gas reservoirs. J. Can. Petrol. Technol. 45 (12).

21. Friedel, T., 2004. Numerical Simulation of Production from Tight-Gas Reservoirs by Advanced Stimulation Technologies (Ph.D. thesis). Freiberg University.

22. Gataullin, T.I., 2008. Modeling of hydraulically fractured wells in full field reservoir simulation model. In: SPE 117421 Presented at the 2008 SPE Russian Oil & Gas Technical Conference & Exhibition, Moscow, Russia, 28–30 Oct.

23. Hajibeygi, H., Karvounis, D., Jenny, P., 2011. A hierarchical fracture model for the iterative multiscale finite volume method. J. Comput. Phys. 230, 8729–8743.

24. Ibrahim, M., 2013. Development of new well index equation for fracture wells. In: SPE 164017 Presented at the SPE Middle East Unconventional Gas Conference and Exhibition, Muscat, Oman, 28–30 Jan.

25. Lefevre, D., Pellissier, G., Sabathier, J.C., 1993. A new reservoir simulation system for a better reservoir management. In: SPE 25604 Presented at the Middle East Oil Show, Bahrain, 3–6 April.

26. Li, L., Lee, S.H., 2008. Efficient field-scale simulation of black oil in a naturally fractured reservoir through discrete fracture networks and homogenized media. SPE Reserv. Eval. Eng. 11 (4), 750–758.

27. Mascarenhas, O., Durlofsky, L.J., 2000. Coarse scale simulation of horizontal wells in heterogeneous reservoirs. J. Petrol. Sci. Eng. 25, 135–147.

28. Medeiros, F., Ozkan, E., Kazemi, H., 2007. Productivity and drainage area of fractured horizontal wells in tight gas reservoirs. In: SPE 108110 Presented at the 2007 Rocky Mountain Oil & Gas Technical Symposium, Denver, CO, USA, 16–18 April.

29. Moinfar, A., Varavei, A., Sepehrnoori, K., Russel, T.J., 2013. Development of a coupled dual continuum and discrete fracture model for the simulation of unconventional reservoirs. In: Paper SPE 163647 Presented at the SPE Reservoir Simulation Symposium, The Woodlands, TX, 18–20 Feb.

30. Nedelec, J.C., 2001. Acoustic and electromagnetic equations: integral representations for harmonic problems. Applied Mathematical Sciences, vol. 144. Springer.

31. Norbeck, J., Huang, H., Podgorney, R., Horne, R., 2014. An integrated discrete fracture model for description of dynamic behavior in fractured reservoirs. In: Proceedings of the Thirty-Ninth Workshop on Geothermal Reservoir Engineering Stanford University, Stanford, CA, February 24–26.

32. Peaceman, D.W., 1978. Interpretation of wellblock pressures in numerical reservoir simulation. In: Proceedings of the SPEJ, June. Trans. AIME, 253, 183–194.

33. Peaceman, D.W., 1983. Interpretation of wellblock pressures in numerical reservoir simulation with nonsquare gridblocks and anisotropic permeability. In: Proceedings of the SPEJ, June, pp. 531.

34. Peaceman, D.W., 1993. Representation of a horizontal well in numerical reservoir simulation. SPE Adv. Technol. Ser. 1 (1), 7–16.

35. Sadrpanah, H., Charles, T., Fulton, J., 2006. Explicit simulation of multiple hydraulic fractures in horizontal wells. In: SPE 99575 Presented at the SPE Europec/EAGE Annual Conference and Exhibition, Vienna, Austria, 12–15 June.

36. Wu, Y.S., 2002. Numerical simulation of single-phase and multiphase non-Darcy flow in porous and fractured reservoirs. Transp. Porous Media 49 (2), 209–240.

37. Wu, Y.S., Li, J., Ding, D.Y., Wang, C., Di, Y., 2013. A generalized framework model for simulation of gas production in unconventional gas reservoirs. In: SPE 163609, Presented at the SPE Reservoir Simulation Symposium, The Woodlands, TX, 18– 20 Feb.

38. Zhou, Y., King, M.J., 2011. Improved upscaling for flow simulation of tight gas reservoir models. In: SPE 147355 Presented at the SPE ATC&E, Denver, CO, USA, 30 Oct.–2 Nov.

Chapter 8

Surface Energy and Wetting Behavior of Reservoir Rocks

Naveed Arsalana, Jan J. Buitingb, and Quoc P. Nguyena

aDepartment of Petroleum and Geosystems Engineering, The University of Texas at Austin, 1 University Station, C0300, Austin, TX 78712-1061, USA

bSaudi Aramco, Dhahran 31311, Saudi Arabia

ABSTRACT

An accurate description of the surface chemistry of the reservoir rock–fluid system is essential to understand the attractive forces between the various phases (crudes, brines and the rock surface). These physico-chemical interactions determine the fundamental nature of the reservoir wettability and the wetting behavior of fluids on the reservoir rock surface. Inverse gas chromatography (IGC) is used to characterize the surface chemistry of a Saudi Arabian reservoir rock (henceforth referred to as 'reservoir rock') at different moisture coverage and temperatures. This information combined with the surface tension of the interacting reservoir fluids is utilized to develop a new method for quantifying wettability in terms of a wettability index. This index is based on the relative magnitude of the work of adhesion between the rock surface and the competing oleic/aqueous phase.

INTRODUCTION

Since the early days of the petroleum industry, attempts have been made to understand the spreading behavior of reservoir fluids on the rock surface and use this knowledge to improve the oil recovery from the reservoir. This led to the concept of wettability, which describes the tendency of a fluid to spread on a rock surface in the presence of another immiscible fluid. Therefore the reservoirs were usually classified as oil-wet, water-wet or intermediate-wet based on the affinity of the rock surface toward oil or water phase. Wettability assumes significance since it determines fluid distribution in the reservoir and the capillary forces holding them and thus affecting reservoir production, waterflood recovery and the performance of enhanced oil recovery (EOR) processes [1], [2], [3], [4], [5], [6] and [7]. However attempts to describe or generalize the concept of wettability have largely remained unsuccessful [8].

Currently wettability is estimated in the laboratory by restoring the in situ wettability of core samples by aging them at elevated temperatures for long periods of time. There are two standard analyses adopted by the industry for wettability estimation: Amott test and USBM method [2]. Both analyses are rather time consuming and expensive. Thus the ensuing paper is an attempt to develop a fast and reliable alternative technique for wettability estimation utilizing our knowledge of interfacial interactions between the various phases. It has been commonly agreed, that the two major factors affecting wettability are surface morphology and the intermolecular surface forces between the 3 phases (rock–oil–

brine) [1], [8] and [9]. Regardless of the morphology, the wettability of the system is determined by the relative magnitude of the forces of interaction between the two liquid phases and the rock surface [8] and [9]. These fundamental interactions (or surface energies) are usually classified into two classes: Lifshitz–van der Waals interactions (non-polar) and acid–base interactions (polar) [10].

Surface free energy (also called surface energy) is an important thermodynamic characteristic of a solid and is defined as the energy required to form (or increase) the surface by a unit surface under reversible conditions. There are two indirect methods commonly used to assess the surface energy of solids: vapor adsorption measurements using probe vapors and wetting (contact angle) measurements using probe liquids. Contact angle measurement is generally limited in its application to low energy smooth surfaces where finite contact angles can be formed using appropriate probe liquids. In case of irregular particulate materials, wicking measurements are used to infer contact angles. Since many high energy surfaces of interest such as minerals are wet by most liquids, the 'two-liquid' approach is used to obtain finite contact angles for the solid–liquid interface. In contrast the vapor adsorption measurements using inverse gas chromatography (IGC) at infinite dilution involves studying the individual interaction of the probe molecules with the surface sites. This approach enables an accurate description of the surface at different temperatures and other physical conditions by taking into account surface heterogeneity and the interaction forces responsible for the adsorption. A brief review of the technique and application of IGC has been can be found in the literature [11] and [12].

Thus the focus of our study is to quantify and understand the nature of these interactions by using inverse gas chromatography and use this knowledge to determine the wettability of the rock surface. The authors have successfully demonstrated the technique to quantify these fundamental interactions by characterizing the surface energetics of some sandstone and carbonate rocks using inverse gas chromatography [13] and [14]. Here we extend this technique to a carbonate rock obtained from a Saudi Arabian reservoir. This information is used it to demonstrate a new approach to quantify the wettability of a reservoir rock by relating it to a wettability index. The method for calculating the wettability index of the reservoir rock is based on measuring the difference between the work of adhesion between the two liquid phases and the rock surface using the van Oss–Chaudhury–Good approach. The ensuing paper will illustrate the mechanics of this process right from performing the surface energetic analysis of the reservoir rocks to calculating their respective wettability indices.

THEORY

The principle and technique behind IGC measurements has been extensively discussed in the literature[13] and [14]. In this section, we will describe the process for determining the wettability index for a reservoir rock, in contact with a brine phase and an oil phase. This step utilizes the knowledge of the surface energies of the three interacting phases (rock, oil and brine) in terms of their Lifshitz–van der Waals components and polar components.

The work of adhesion (W^A) is a thermodynamic property and is defined as the work required for separating two different surfaces (denoted by 1 and 2) from each other. In other words, the work of adhesion between any two surfaces determines how strongly the surfaces are attracted to one another.

$$6WA = \gamma_1 + \gamma_2 - \gamma_{12} \tag{1}$$

where γ_{12} is the interfacial tension between the two surfaces 1 and 2, γ_1 is the surface tension of the surface 1 and γ_2 is the surface tension of the surface 2.

Building on the work of Good and Girifalco [15] and Fowkes [10], the van–Oss–Chaudhury–Good model[16] and [17] expresses the work of adhesion W_{12}^A between two surfaces (1 and 2) as follows:

$$W_{12}^A = 2\sqrt{\gamma_1^{LW}\gamma_2^{LW}} + 2\sqrt{\gamma_1^-\gamma_2^+} + 2\sqrt{\gamma_1^+\gamma_2^-} \tag{2}$$

where Y_1^{LW} is the Lifshitz–van der Waals component of surface energy of surface 1, Y_1^- is the basic component of surface energy of surface 1 and Y_1^+ is the acidic component of surface energy of surface 1. Similarly Y_2^{LW}, Y_2^- and Y_2^+ represent the Lifshitz–van der Waals component, basic component and acidic component of surface energy of surface 2 respectively. Knowing the surface energy and its components for all the three phases: rock (S), brine (W) and oil (O), we can calculate the work of adhesion W_{WS}^A between brine and the rock surface using Eq. (2) as follows:

$$W_{WS}^A = 2\sqrt{\gamma_W^{LW}\gamma_S^{LW}} + 2\sqrt{\gamma_W^-\gamma_S^+} + 2\sqrt{\gamma_W^+\gamma_S^-} \tag{3}$$

where Y_S^{LW} is the Lifshitz–van der Waals component of surface energy of reservoir rock surface, Y_S^- is the basic component of surface energy of reservoir rock surface and Y_S^+ is the acidic component of surface energy of reservoir rock surface. Similarly Y_W^{LW}, Y_W^- and Y_W^+ represent the Lifshitz–van der Waals component, basic component and acidic component of surface tension of brine respectively. Similarly the work of adhesion W_{OS}^A between the oil and the rock surface can be calculated as follows:

$$W_{OS}^A = 2\sqrt{\gamma_O^{LW}\gamma_S^{LW}} + 2\sqrt{\gamma_O^-\gamma_S^+} + 2\sqrt{\gamma_O^+\gamma_S^-} \tag{4}$$

where Y_O^{LW}, Y_O^- and Y_O^- represent the Lifshitz–van der Waals component, basic component and acidic component of surface tension of oil respectively. Based on our hypothesis stated at the beginning, the wettability of the system is determined by the relative magnitude of the forces of interaction between the two liquid phases and the rock surface. The relative wetting property (Δ_W) is defined as follows:

$$\Delta_W = W^A_{WS} - W^A_{OS}$$

(5)

If Δ_W is positive, this implies water preferentially wets the rock surface, whereas if Δ_W is negative, this implies oil preferentially wets the rock surface. A value of zero for Δ_W indicates that there is no preferential interaction between the rock surface to either the oil phase or the brine phase. Thus the rock surface behaves like an intermediate wet surface if $\Delta_W = 0$. Based on this approach, we propose a wettability index (WI) by normalizing the relative wetting property (Δ_W) as follows:

$$WI = \frac{W^A_{WS} - W^A_{OS}}{W^A_{WS} + W^A_{OS}}$$

(6)

Thus for a water-wet rock, the value of WI scales from 0 (intermediate wet) to 1 (strongly water-wet), whereas for an oil-wet rock, the value of WI scales from 0 (intermediate wet) to -1 (strongly oil-wet). Therefore we have formulated a simple and accurate scale for mapping the wettability index between 1 and -1.

MATERIALS AND APPARATUS

The reservoir rock samples were ground using mortar and pestle and were subsequently sieved to obtain 100 mesh sieve fraction. The HPLC grade polar (dichloromethane and ethyl acetate) and non-polar (C_5–C_9 n-alkanes) solvents used for chromatographic injection were obtained from Acros Organics.

EXPERIMENTAL PROCEDURE

Characterization of Surface Energy of the Reservoir Rock with Increasing Moisture Coverage

The inverse gas chromatograph used in our study was built by Surface Measurement Systems Ltd (iGC-2000 model). The setup utilizes a series of mass flow controllers to prepare mixtures of helium carrier gas and probe solvents (non-polar and polar solvents). An automated injection valve injects 250 µL of the elution mixture into the carrier gas flowing through the column into the detectors. A thermal conductivity detector (TCD) and flame ionization detector (FID) are coupled together at the end of the column for the sensitive analysis of the probe molecules. The chromatographic column used for holding the powdered mineral sample is housed in a separate column oven to maintain it at a constant temperature. Silanized glass wool (Aldrich) is used to pack the powdered rock samples in place. The silanized glass columns measured 30 cm in length and had the following dimensions (6 mm o.d and 4 mm i.d).

The sieved rock sample was washed with ethanol and dried in the oven at 150 °C for nearly 30 min. The cleaned sample was packed in a column and flushed with nitrogen gas at 150 °C for over 5 h to minimize any further moisture contamination. Before injecting the probe solvents, the column is further conditioned with helium gas at the test temperature and relative humidity for 2 h each. Moisture was deposited on the rock surface by controlling the relative humidity of the carrier gas. For studying a dry surface, the carrier gas had zero relative humidity. First moment method was employed to deduce the retention times from the elution curves generated by the detectors.

RESULTS AND DISCUSSION

Characterization of Surface Energy of the Reservoir Rock with Increasing Moisture Coverage

Water Adsorption Isotherms

The water adsorption isotherms for the reservoir rock at 30 and 50 °C have been plotted in Fig. 1. Using N_2BET adsorption analysis, the specific surface area of the reservoir rock was measured to be 0.5584 m^2/g. The water adsorption isotherm displays a strong type II isotherm behavior (Fig. 1). This indicates the formation of multilayers at higher RH. The calculated monolayer coverage is achieved at 20% RH and beyond 70% RH, moisture is deposited on the surface in a multilayered fashion. For a type II mechanism, the heat of adsorption is much higher than the heat of condensation i.e. the molecules would rather interact with the surface than with each other.

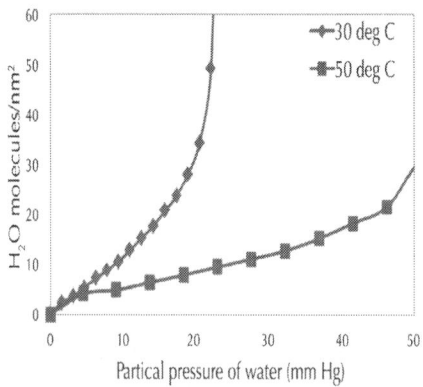

Figure 1: Water adsorption isotherms for the reservoir rock at 30 and 50 °C.

Surface Interactions or Components of Surface Energy

The total surface free energy of a reservoir rock comprises of Lifshitz–van der Waals, acidic and basic components. As the water adsorption isotherms for the reservoir rock (Fig. 1) have indicated, we observe increasing water surface coverage with increase in the relative humidity of the carrier gas stream. This results in a corresponding reduction in the surface energy of the reservoir rock with increase in moisture deposition onto the surface. The mineral surfaces are usually strongly heterogeneous and have a high surface free energy. The adsorbed water layer stabilizes the surface by presenting a lower energy homogeneous surface to the probe molecules to interact [18]. Thus the total surface energy of the reservoir rock decreases sharply as moisture content is slowly increased and eventually attains a plateau at higher relative humidity (Fig. 2). The slight increase in total surface energy at greater relative humidity at 50 °C is indicative of the effect exerted by the solubility of the probe molecules in the adsorbed water layers, which will be discussed in the coming paragraphs.

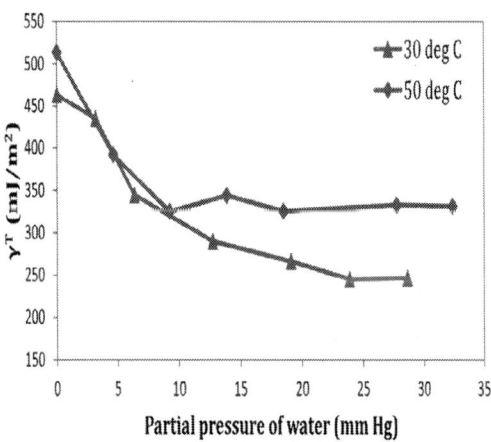

Figure 2: Effect of water coverage on the total component of surface energy for the reservoir rock at 30 and 50 °C.

The Lifshitz–van der Waals component of surface energy comprises of the following interactions: Keesom (dipole–dipole interactions), Debye (dipole-induced–dipole interactions) and London dispersion forces (induced dipole–induced dipole interactions). Similar to the behavior of the total surface energy, we observe with increase in moisture coverage, the Lifshitz–van der Waals component of surface energy decreases rapidly and at high relative humidity will attain a plateau as shown in Fig. 3. This decrease is most rapid at low water coverage since the adsorbate molecules will prefer to occupy the most energetic sites on the reservoir rock. Thus the heat of adsorption of the first layer is dependent on the water coverage. For the second and subsequent layers, this is not expected to be significant as they are occurring on a layer of adsorbed water molecules. Thus at greater water coverage, the decrease of the Lifshitz–van der Waals component of surface energy begins to plateau.

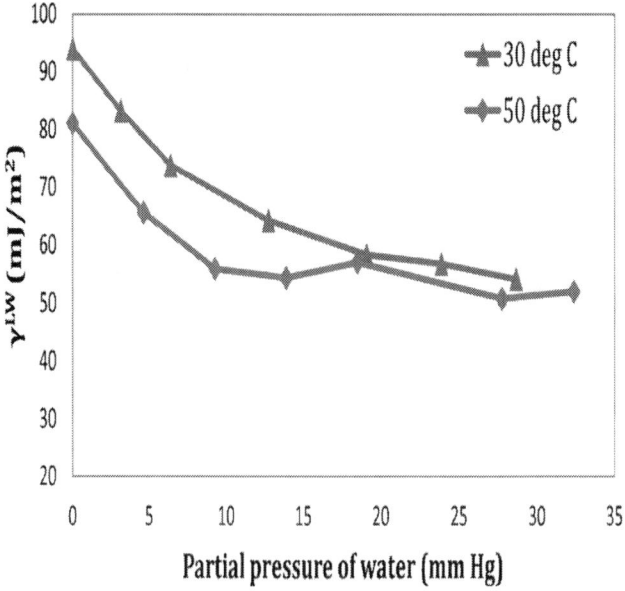

Figure 3: Effect of water coverage on Lifshitz–van der Waals component of surface energy for the reservoir rock at 30 and 50 °C.

Ideally at higher RH, it is expected that the Lifshitz–van der Waals component of surface energy of the moisture covered rock surface should tend toward the Lifshitz–van der Waals component of surface tension of bulk water. However the higher plateau of the Lifshitz–van der Waals component of surface energy at high moisture coverage indicates non-uniform distribution of water layers, leaving bare mineral surfaces to interact with the probe solvents [19]. This non-uniformity in the water coverage of the surface sites may be induced because of imperfections on the surface [20], which may be created when the mineral sample is ground.

The behavior of acid–base components of surface energy of the reservoir rock can be better understood in the light of the following reactions that take place on account of physisorption and chemisorption of water on the surface. At any given point, the surface of a mineral is covered with both physisorbed and chemisorbed water as soon as it comes in contact with moisture. Since the reservoir rock is a carbonate rock, the surface is usually populated by calcium, magnesium and carbonate groups. During physisorption (associative adsorption), the O^{2-} atom associated with water covers the stronger acidic surface sites (Ca^{2+}, Mg^{2+}), while exposing the weakly polar H^+ of the water molecule on the exterior. Similarly the stronger basic surface site (CO_3^{2-}) is covered by the H^+ of the water molecule while exposing the weakly polar O^{2-} site on the exterior. In this way, the stronger polar sites due to the reservoir rock surface are replaced by the weaker polar sites of the water layer. This leads to a decrease in acidic and basic components of surface energy with increasing relative humidity (Fig. 4 and Fig. 5). Chemisorption (dissociative adsorption) is preceded by the decomposition of $CaCO_3$ to CaO, which reacts with water to result in the formation of isolated hydroxyl groups at the Ca^{2+} sites and surface bicarbonate anions. In our study, we assume chemisorption is always present even in a dry test rock surface.

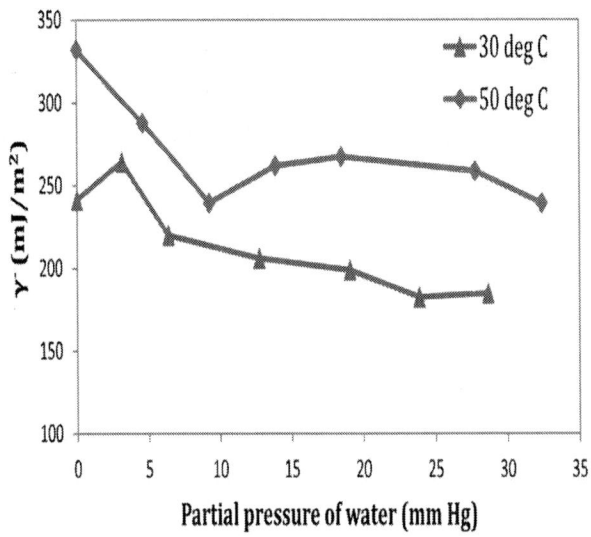

Figure 4: Effect of water coverage on the basic component of surface energy for the reservoir rock at 30 and 50 °C.

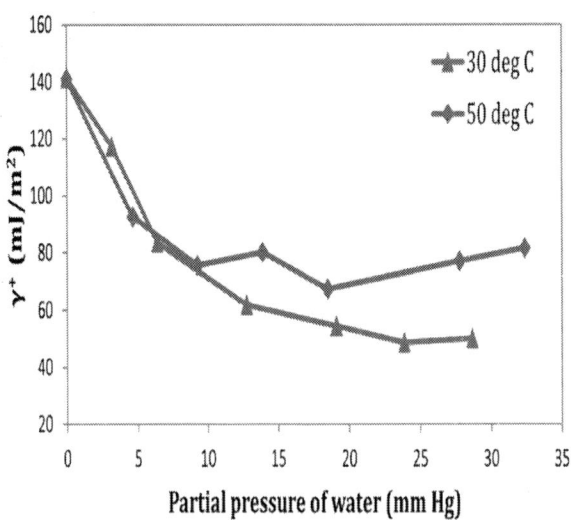

Figure 5: Effect of water coverage on the acidic component of surface energy for the reservoir rock at 30 and 50 °C.

In a remarkable observation, we see that there is a sudden increase in the plateau of the acidic and basic components of surface energy (Fig. 4 and Fig. 5) at greater water coverage at 50 °C. This can be explained due to increased solubility of the polar solvents in the water multilayers deposited on the surface of the mineral at high relative humidity. This causes an artificial increase in the retention times of the probe solvents which distorts surface energy measurements. In comparison, no such similar effect is observed in case of the Lifshitz–van der Waals component of surface energy because of the insoluble nature of the non-polar solvents in water.

Since we observe appreciably large acidic and basic components of surface energy, we conclude that the reservoir rock surface is amphoteric in nature. This implies that the dynamic wetting properties are determined by both the reservoir surface and reservoir fluid chemistry. The reservoir fluids are usually a diverse mixture of crude oils and brines. Consequently, both crude oil and brine exhibit different polar and non-polar components of surface energy/tension. In line with our hypothesis, it is the relative strength of the interactions (measured by work of adhesion) between the crude oil–rock surface and brine–rock surface that would determine whether the rock behaves as an oil-wet or water-wet rock.

Work of Adhesion and the Wetting Behavior of Different Polar and Non-Polar Liquids

Since we know the surface tension data for some common polar and non-polar liquids, we will study their wetting behavior with respect to the reservoir rock and water phase. It is commonly assumed that reservoir oils are non-polar in nature and thus interact only by Lifshitz–van der Waals interactions. In the absence of surface tension data with their polar components for reservoir oils, we approximated the surface tension data for pure species (decane, hexadecane, chloroform, and toluene) from the literature [11] to be that of paraffinic oil, slightly acidic oil and slightly basic oil. The wetting behavior of these pure species will give us an idea as to how these types of oil will behave on the reservoir rock surface.

Using the surface tension data provided in Table 1 into Eq. (3) and (4), we calculated the work of adhesion between the different liquid phases and reservoir rock surface. These calculated works of adhesion are depicted pictorially in Fig. 6. Using this information, we also computed the relative wetting parameters and wettability indices for the reservoir rock with respect to the two liquid phases. This has been displayed inTable 2.

Table 1: Surface tension and its components at 20 C [11]

Liquid	Nature	γ^{LW}	γ^-	γ^+	γ^T
Decane (D)	Paraffinic oil	23.83	0	0	23.83
Hexadecane (H)	Paraffinic oil	27.47	0	0	27.47
Chloroform (CF)	Acidic oil	27.2	0	3.8	27.2
Toluene (T)	Basic oil	28.5	2.3	0	28.5
Water (W)	Brine	21.8	25.5	25.5	71.8

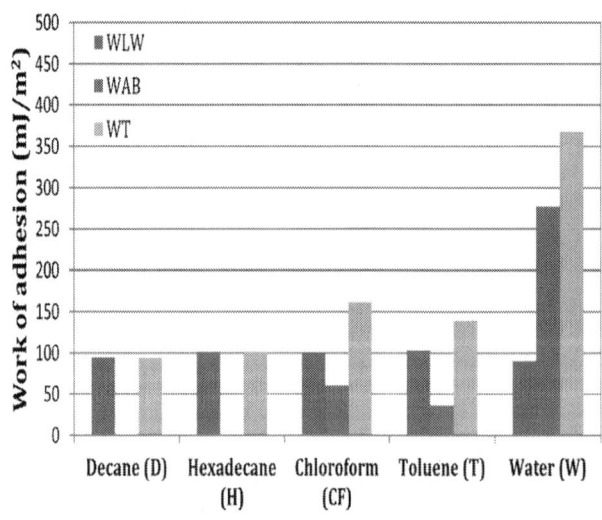

Figure 6: Work of adhesion for the different wetting fluids on the reservoir rock.

Table 2: Relative wetting parameter and wettability index for the reservoir rock with respect to different fluid combinations.

Phase 1	Phase 2	Δ_w	WI
Decane (D)	Water (W)	272.918	0.59
Hexadecane (H)	Water (W)	265.947	0.57
Chloroform (CF)	Water (W)	205.881	0.39
Toluene (T)	Water (W)	227.977	0.45

In case of paraffinic oils such as decane and hexadecane, the work of adhesion with the reservoir surface is only due to Lifshitz–van der Waals interactions. On the contrary, water due to its polar nature shows very high work of adhesion. Thus water wets the reservoir rock surface quite significantly in comparison to the paraffinic fluids which is confirmed by the higher WI (0.59 and 0.57) in Table 2. Chloroform and Toluene show slightly higher works on adhesion due to their slightly monopolar nature. In comparison, water again shows significantly higher works of adhesion causing the rock surface to act as a water-wet surface. Thus due to increased interaction of the oleic phase (chloroform and toluene, due to their monopolar character) with respect to the water phase, the wettability indices scale back in the 0–1 water wetness scale (0.39 and 0.45).

Bitumen–Reservoir Rock Interactions and the Wettability Index

Let us consider the interactions of a typical mixture of oil such as bitumen with the reservoir rock surface, for which the surface energy was determined using inverse gas chromatography [21]. The surface energy and its components for bitumen and water are listed in Table 3. We observe that bitumen displays mostly Lifshitz–van der Waals component of surface energy and very small polarity compared to the reservoir rock surface.

Table 3: Surface tension and its components for bitumen and water at 30 C [11] and [21]

Liquid	γ^{LW}	γ^-	γ^+	γ^T
Bitumen (B)	48.3	0.4	0.8	50.4
Water (W)	21.8	25.5	25.5	71.8

Using the approach outlined in Section 2, the work of adhesion between the two liquid phases with the reservoir rock surface is calculated and tabulated in Table 4. The work of adhesion between bitumen and the reservoir rock surface is dominated by Lifshitz–van der Waals interactions. Similar to what we observed in Section 5.2, the polar nature of water causes a stronger interaction between the water phase and the reservoir rock surface. Thus water preferentially wets the reservoir rock surface in comparison to bitumen as indicated by the positive $_w$ and WI (0.35) for the reservoir rock surface as displayed in Table 5. Similarly one notices, the reservoir rock in the bitumen–reservoir rock–water system (WI = 0.35) behaves as a less water-wet surface when compared to the paraffinic oil–reservoir rock–water system (WI = 0.57–0.59). Thereby also indicating the role of surface properties of the fluids on the wetting behavior of the rock surface.

Table 4: Work of adhesion between the liquid (bitumen–water) and the reservoir rock surface

	Bitumen (B)	Water (W)
W^{LW}	134.73	90.52
W^{AB}	42.84	277.04
W^T	177.57	367.55

Table 5: Relative wetting parameter and wettability index for the reservoir rock

	Relative wetting, Δ_w	Wettability index, WI
Reservoir rock	189.99	0.35

Based on the scale we developed, +1 corresponds to very strongly water-wet rock while −1 corresponds to very strongly oil-wet rock. Thus the reservoir rock surface studied in comparison to their adhesion strength with bitumen and water phase indicates that they are fairly water-wet due to a positive WI value lying in the range of 0–1.

CONCLUSIONS

In this study, we have introduced the technique of inverse gas chromatography to characterize the surface energetics of a reservoir rock at varying moisture coverage and at 30 and 50 °C. Using this approach, the polar and non-polar components of the surface energy for the reservoir rock were mapped and quantified at varying surface conditions. The Lifshitz–van der Waals and acid base component of surface energy showed a decreasing trend with increase in moisture coverage due to water (low energy surface in comparison to the rock) preferentially clustering around the high energy sites. The surface chemistry of the reservoir rock was further characterized using water adsorption isotherms.

The surface energies for the reservoir rock and various polar and non-polar fluids were used to demonstrate a new approach to quantify the wettability of a reservoir rock by relating it to a wettability index. The method for calculating the wettability index of the reservoir rock was based on the hypothesis that the wettability of the system is determined by the relative magnitude of the forces of interaction between the two competing liquid phases and the reservoir rock surface. The work of adhesion and the wetting behavior of various pure fluids (decane, hexadecane, chloroform, toluene and water) against the reservoir rock surface were calculated using the van Oss–Chaudhury–Good approach. The wetting behavior of these fluids was treated as approximations of the wetting behavior paraffinic–acidic–basic oils on the reservoir rock surface. Based on the relative wetting behavior of the two competing phases against the reservoir surface, the wettability index for the mineral surface was proposed and calculated. The scale ranges from −1 to +1, with −1 referring to very strongly oil-wet rock while +1 refers to very strongly water rock.

Finally this approach was validated against the reservoir conditions by treating bitumen and water as the two competing fluid phases and computing the wettability index for the reservoir rock surface. The wettability index of 0.35 indicated that reservoir rock shows a predominantly water-wet behavior in relation to bitumen and the water phase.

Acknowledgement We would like to thank Saudi Aramco and FCMG for their supports.

REFERENCES

1. W.O. Anderson, Wettability literature survey – Part 1: rock/oil/brine interactions and the effects of core handling on wettability, J. Pet. Technol. (1986) 1125–1149.

2. W.O. Anderson, Wettability literature survey – Part 2: wettability measurement, J. Pet. Technol. (1986) 1246–1262.

3. W.O. Anderson, Wettability literature survey – Part 3: the effects of wettability on the electrical properties of porous media, J. Pet. Technol. (1986) 1371–1378.

4. W.O. Anderson, Wettability literature survey – Part 4: the effects of wettability on capillary pressure, J. Pet. Technol. (1987) 1283–1300.

5. W.O. Anderson, Wettability literature survey – Part 5: the effects of wettability on relative permeability, J. Pet. Technol. (1987) 1453–1468.

6. W.O. Anderson, Wettability literature survey – Part 6: the effects of wettability on waterflooding, J. Pet. Technol. (1987) 1605–1622.

7. N.R. Morrow, Wettability and its effect on oil recovery, SPE Distinguished Author series, J. Pet. Technol. 42 (1990) 1476–1484.

8. C. Drummond, J. Israelachvili, Surface forces and wettability, J. Pet. Sci. Eng. 33 (2002) 123–133.

9. G.J. Hirasaki, Wettability: fundamentals and surface forces, Soc. Pet. Eng. Form. Eval. (1991) 217–226.

10. F.M. Fowkes, Attractive forces at interfaces, Ind. Eng. Chem. 12 (1964) 40–52.

11. F.M. Etzler, Characterization of surface free energies and surface chemistry of solids Contact Angle, Wettability and Adhesion, vol. 3, 2003, pp. 1–46.

12. A. Voelkel, B. Strzemiecka, K. Adamska, K. Milczewska, Inverse gas chromatography as a source of physiochemical data, J. Chromatogr. A 1216 (2009) 1551–1566.

13. N. Arsalan, S.S. Palayangoda, D.J. Burnett, J.J. Buiting, Q.P. Nguyen, Surface energy characterization of sandstone rocks, J. Phys. Chem. Solids 74 (2013) 1069–1077.

14. N. Arsalan, S.S. Palayangoda, D.J. Burnett, J.J. Buiting, Q.P. Nguyen, Surface energy characterization of carbonate rocks, Colloids Surf. A 436 (2013) 139–147.

15. L.A. Girifalco, R.J. Good, A theory for the estimation of surface and interfacial energies. I Derivation and application to interfacial tension, J. Phys. Chem. C 61 (1957) 904.

16. C.J. van Oss, M.K. Chaudhury, R.J. Good, Monopolar surfaces, Adv. Colloid Interface Sci. 28 (1987) 35 64.

17. C.J. van Oss, M.K. Chaudhury, R.J. Good, Interfacial Lifshitz–van der Waals and polar interactions in macroscopic systems, Chem. Rev. 88 (1988) 927–941.

18. D.S. Keller, P. Luner, Surface energetics of calcium carbonates using inverse gas chromatography, Colloids Surf. A 161 (2000) 41–415.

19. P.K. Mogili, P.D. Kleiber, M.A. Young, V.H. Grassian, Heterogeneous uptake of ozone on reactive components of mineral dust aerosol: an environmental aerosol reaction chamber study, J. Phys. Chem. A 110 (2006) 13799– 13807.

20. A. Rahaman, V.H. Grassian, C.J. Margulis, Dynamics of water adsorption onto calcite surface as a function of relative humidity, J. Phys. Chem. C 112 (2008) 2109–2115.

21. A.W. Hefer, D.N. Little, B.E. Herbert, Bitumen surface energy characterization by inverse gas chromatography, J. Test. Eval. 35 (2007) 1–8.

Evolving an Accurate Model Based on Machine Learning Approach for Prediction of Dew-Point Pressure in Gas Condensate Reservoirs

Seyed Mohammad Javad Majidi[a], Amin Shokrollahi[a], Milad Arabloo[a], Ramin Mahdikhani-Soleymanloo[b], and Mohsen Masihi[a]

[a]Chemical & Petroleum Engineering Department, Sharif University of Technology, Azadi Ave., Tehran, Iran

[b]Petroleum Engineering Department, Petroleum University of Technology, Ahwaz, Iran

ABSTRACT

Over the years, accurate prediction of dew-point pressure of gas condensate has been a vital importance in reservoir evaluation. Although various scientists and researchers have proposed correlations for this purpose since 1942, but most of these models fail to provide the desired accuracy in prediction of dew-point pressure. Therefore, further improvement is still needed. The objective of this study is to present an improved artificial neural network (ANN) method to predict dew-point pressures in gas condensate reservoirs. The model was developed and tested using a total set of 562 experimental data point from different gas condensate fluids covering a wide range of variables. After a series of optimization processes by monitoring the networks performance, the best network structure was selected. This study also presents a detailed comparison between the results predicted by this ANN model and those of other universal empirical correlations for estimation dew-point pressure. The results showed that the developed model outperforms all the existing methods and provides predictions in acceptable agreement with experimental data. Also it is shown that the improved ANN model is capable of simulating the actual physical trend of the dew-point pressure versus temperature between the cricondenbar and cricondenterm on the phase envelope. Finally, an outlier diagnosis was performed on the whole data set to detect the erroneous measurements from experimental data.

INTRODUCTION

Well deliverability in gas condensate reservoirs, is declined rapidly by condensate banking when the flowing bottomhole pressure falls below the dew point pressure. The liquid dropout around wellbore causes the productivity decreases. This ring of increased condensation saturation around the wellbore reduces effective permeability to gas and results in rapid well-productivity decline (Elsharkawy, 2002). Therefore accurate prediction of dew-point pressure of gas condensate reservoirs is a critical element in planning the development of gas condensate reservoirs.

Dew-point pressure of a gas condensate sample is determined through the constant mass expansion test. Although the experimentally measurement of dew-point pressure is very accurate and reliable, but it is very time consuming and costly process for gas condensate reservoirs (Grieves and Thodos, 1963, Pedersen et al., 1988 and Sage and Olds, 1947).

There are generally two methods to estimate dew-point pressure of gas condensate reservoirs. The first method uses the equations of state and the second one uses empirical correlations. Equations of state do not have ability to simulate the phase behaviour of light oil and gas condensate reservoirs, especially in the retrograde area. Besides, all the developed empirical correlations have not enough accuracy (Shokir, 2008). Therefore, there is a need to methods for predicting the dew-point pressure of gas condensate reservoirs accurately.

In 1942, Kurata and Katz (1942) developed a correlation to estimate critical properties of light hydrocarbon mixtures. They used a limited number of dew-point pressure to develop their correlation, but the effect of fluid composition were neglected.

Olds et al. (1945) used the compositions of oil and gas samples obtained from primary separator of a well in the Paloma field, to develop a new correlation for prediction of dew-point pressure of gas condensate reservoirs in graphical and tabular form. They also investigated the effect of omission of intermediate molecular weight on dew-point pressure. They showed that intermediate molecular weight components have a considerable effect on dew-point pressure. Olds et al. (1949) experimentally investigated the volumetric behaviour for different mixtures of gas condensate reservoirs from San Joaquin Valley field. They presented a correlation which relates dew-point pressure to stock tank API oil gravity, gas–oil ratio, and temperature in tabulated and graphical forms.

Reamer and Sage (1950) used five different pairs of fluids from a field in Louisiana. They studied the effects of temperature and gas–oil ratio on dew-point pressure. The results are presented in graphical forms. Moreover, they concluded that due to complex effect of fluid composition on dew-point pressure, a general correlation for this purpose is not possible.

Organick and Golding (1952) introduced a simple correlation in the form of working charts for prediction of dew-point pressure in gas condensate reservoirs. Nemeth and Kennedy (1967) proposed a mathematical relationship between the dew-point pressure of a hydrocarbon fluid and its composition, temperature and characteristics of the C_{7+} fraction. They used 579 data sets from 480 hydrocarbon systems. Their correlation contains 11 coefficients with an average deviation of 7.4%.

Later, Crogh (1996) improved the Nemeth and Kennedy correlation for better prediction of dew-point pressure. One should note here that, reservoir temperature was not considered in the developed correlation.

Humoud and Al-Marhoun (2001) developed an empirical correlation to estimate dew-point pressure in gas condensate systems using available field data. They used different gas condensate fluid samples in Middle East. The newly proposed correlation predicts with an average absolute error of 4.3%, and a maximum relative error of 15.1%.

Elsharkawy (2002) published another mathematical model to predict dew-point pressure of gas condensate reservoirs using parameters involved in Nemeth and Kennedy correlation. Their correlation contains 19 constants with an average absolute deviation of 7.68%.

Shokir (2008) developed another new mathematical genetic programming based model. In the proposed model the effect of SG_{C7+} was not considered. The comparison indicated that the developed model is more accurate than Elsharkawy (2002) correlation and Peng–Robinson EOS. Although numerous number of equations of state (Martin, 1979, Soave, 1972 and Zudkevitch and Joffe, 1970), have been developed to model reservoir fluid phase behaviour, but they could not simulate the phase behaviour of complex hydrocarbons such as gas condensates in the retrograde region (Sarkar et al., 1991).

Nowadays, with the aid of computers, artificial intelligence has become an inseparable part of engineering predictions especially in chemical and petroleum engineering (Ahmadi et al., 2013, Ali Ahmadi et al., 2013,Arabloo et al., 2013, Chamkalani et al., 2013, Farasat et al., 2013, Hemmati-Sarapardeh et al., 2013,Kumar, 2009, Morooka et al., 2001, Rafiee-Taghanaki et al., 2013, Roosta et al., 2011, Shokrollahi et al., 2013 and Zendehboudi et al., 2012). The complexity, fuzziness and uncertainty existent in addition to non-linear behaviour of most reservoir parameters such as dew-point pressure in gas condensate reservoirs require a powerful tool to overcome these challenges. To this end, objective of this study is to predict dew-point pressure of retrograde gas condensate reservoirs using a wide range of experimental data points by applying artificial neural network. In addition, results of estimated dew-point pressure by developed intelligent model are also compared with predictions of other empirical correlations. Finally, validity and sensitivity of the proposed model is examined against variation in temperature.

ARTIFICIAL NEURAL NETWORKS (ANN)

A detailed description of neural networks is given elsewhere (Haykin, 1994). ANNs imitates the behaviour of biological neurons and learn by trial and error. These methods have large numbers of computational units connected in a massively parallel structure and do not require an explicit formulation of the mathematical or physical relationships of the handled problem (Chouai et al., 2002). The overall computational model consists of a reconfigurable interconnection of simple elements, or units. Fig. 1depicts a sample network, where units are denoted by circles and interconnections are shown as lines. Notice that some units in Fig. 1 interface directly with the outside world, whereas others are "hidden" or internal. Individual units implement a local function, and the overall network of interconnected units displays a corresponding functionality. In other words, the system's knowledge, experience, or training is stored in the form of network interconnections. Analysis of this functionality, except through training and test examples, is often difficult. To be useful, neural systems must be capable of storing information (i.e., they must be "trainable"). Neural systems are trained in the hope that they will subsequently display correct associative behaviour when presented with new patterns to recognize or classify (Schalkoff, 1997).

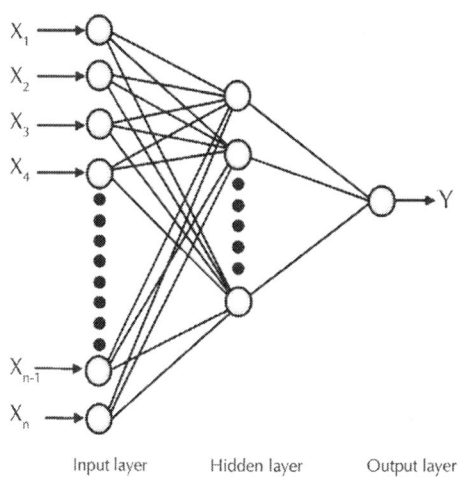

Figure 1: Schematic of a typical network.

To optimize the network performance function, during training process the connection weights and biases are adjusted by a trial and error process. There are many different types of neural networks that differ from each other in network structure and/or learning algorithm. Many parameters must be optimized for designing an acceptable artificial neural network including learning algorithm, number of hidden layers, number of neurons in each layer and transfer functions. There are different learning algorithms that can be applied to train an artificial neural network. One of the most popular and commonly used networks is the multilayer feed forward network (Mohammadi and Richon, 2007, Moradi et al.,

2011, Rój and Wilk, 1998 and Turan et al., 2011). ANN is trained by a set of known inputs and outputs. It learns the patterns of these inputs and outputs by manipulating the weights. The weights are adjusted until the optimization criterion, which is quadratic error between observed data and computed output to be minimized. So in this work the multilayer feed forward network has been used. There are various algorithms to minimize the error during neural network training and obtain the optimum value of network weights and biases such as Levenberg–Marquardt (LM), Resilient back Propagation (RP), and Scaled Conjugate Gradient (SCG). The defect of feed forward neural networks is the determination of the ideal number of neurons in the hidden layer(s); few neurons produce a network with low precision and a higher number leads to over fitting and bad quality of interpolation. The use of techniques such as Levenberg–Marquardt algorithm can help overcome this problem (Levenberg, 1944, Marquardt, 1963 and Wilamowski et al., 2001). In this work the Levenberg–Marquardt algorithm has been used during training process.

ARTIFICIAL NEURAL-NETWORK MODEL DEVELOPMENT

Methodology plan to perform all the tasks involved in this work is as follows.

Data Acquisition and Analysis

One of the most important decisions in developing the artificial neural-network (ANN) model is choosing the experimental data covering widest possible range of conditions under which condensate reservoirs occur.

A total set of 562 experimental data point of constant volume depletion test for different retrograde gas condensate fluids were used to develop ANN model. These data were obtained from open literature (Nemeth and Kennedy, 1967).

Each data set includes experimental values for reservoir temperature, hydrocarbon composition of C_1 through C_{7+}, non-hydrocarbon composition (N_2, CO_2, H_2S), specific gravity of the heptane plus fraction (SG_{C7+}), molecular weight of the heptane plus fraction (MW_{C7+}), and dew-point pressure. Ranges and their corresponding statistical parameters of the input/output data are presented in Table 1. As it can be seen, the data represent a wide range of conditions. Reservoir temperatures varied from 40 to 320 °F, specific gravities of the heptane plus from 0.733 to 0.8681, molecular weights of the heptane plus from 106 to 235, and dew-point pressures from 1405 to 10,790 psia.

Table 1: Ranges and their corresponding statistical parameters of the input/output data used for constructing ANN model

Parameter	Minimum	Maximum	Average	Standard deviation
Temperature (°F)	40	320	205.15	55
Molecular weight (MW_{C7+})	106	235	148.19	23.15
Specific gravity (SG_{C7+})	0.7330	0.8681	0.7881	0.0241
Dew-point pressures (psia)	1405	10,790	4748.22	1624.26
Methane (mole fraction)	0.0349	0.9668	0.8027	0.1230
Ethane (mole fraction)	0.0037	0.1513	0.0578	0.0300
Propane (mole fraction)	0.0011	0.1090	0.0302	0.0199
Butanes (mole fraction)	0.0017	0.2030	0.0204	0.0162
Pentanes (mole fraction)	0.0006	0.0631	0.0120	0.0109
Hexanes (mole fraction)	0.0004	0.0510	0.0092	0.0080
Heptane-plus (mole fraction)	0.0019	0.1356	0.0370	0.0280
Nitrogen (mole fraction)	0.0000	0.4322	0.0102	0.0317
Carbon dioxide (mole fraction)	0.0000	0.9192	0.0156	0.0565
Hydrogen sulfide (mole fraction)	0.0000	0.2986	0.0066	0.0305

ANN Model Construction

After collecting and screening of the data sets, the next step is to construct the model. To this end, it is essential to select the input variables to feed the ANN model. A review on recent researches (Elsharkawy, 2002, Nemeth and Kennedy, 1967 and Shokir, 2008) shows that dew-point pressure of retrograde gas condensate reservoirs is a function of reservoir temperature (T_{res}), hydrocarbon and non-hydrocarbon reservoir fluid compositions (z_i), and characteristics of the heptane plus fraction (SG_{C7+}, MW_{C7+}) as follows:

$$P_d = f(T_{res.}, z_i, SG_{C_{7+}}, MW_{C_{7+}})$$

$$(1)$$

As illustrated in Fig. 2, the purpose of this study is to predict dew-point pressure of retrograde gas condensate reservoir using 13 input parameters namely reservoir temperature, hydrocarbon and non-hydrocarbon composition (C_1 through C_{7+}, N_2, CO_2, H_2S), specific gravity of the heptane plus fraction (SG_{C7+}), and molecular weight of the heptane plus fraction (MW_{C7+}).

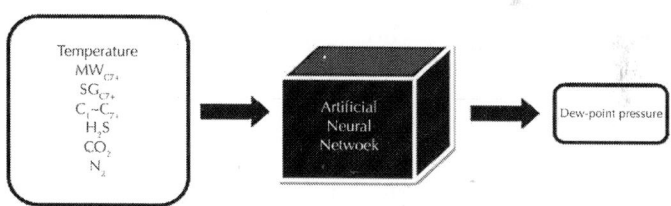

Figure 2: Developed ANN model scheme for prediction of dew point pressure.

In this part of study, the objective is to find the optimal performance of ANN model for prediction of dew-point pressure. First, data sets are randomly split into three partitions (i.e., training, validation, testing). The training data sets are used to perform the training of the network and constructing an internal model. The validation data sets are used to evaluate the quality and generalization of the developed network during the training phase and the testing data sets are used to examine the final performance of the network.

The accuracy of the model is dependent directly on the architecture of the neural network (Haykin, 1994, Mohammadi and Richon, 2007 and Rój and Wilk, 1998). Also, defining the optimal network architecture that simulates the actual behaviour within the experimental data sets is not an easy task. Therefore to achieve this, different network topologies were constructed. In order to avoid overfitting problems, the data set was divided into three subdata sets. For this purpose, 70%, 15%, 15% of the main data set was randomly selected for the training set, validation set, and test set, respectively. Considering the complexity of the relationship between the selected input and output variables, multiple-layer feed forward back propagation networks were utilized. Feed forward neural networks are the most frequently used to representing non-linear functional mapping between inputs and output (Haykin, 1994). The criterion for selection of the best network structure was chosen by monitoring the networks performance through calculating average absolute percent relative error, root mean square error, and correlation coefficient (Appendix A) between network outputs and experimental values for each inspected structure. Table 2shows the statistical analysis of different network structures. It should be mentioned that n-25-10-5 represent an ANN model having three hidden layers, with first, second, and third layers having 25, 10, and 5 neurons, respectively.

Table 2: Tested networks and their statistical parameters

	Network topology	AAPE	RMSE	R
1	n-15	7.7907	12.102	0.9483
2	n-20	6.9529	10.765	0.9483
3	n-30	7.3405	10.644	0.9578
4	n-25	6.9117	10.967	0.9545
5	n-10-5	7.5585	10.756	0.9536
6	n-15-5	8.2174	12.311	0.949
7	n-20-5	7.5794	10.885	0.9566
8	n-25-5	12.4381	18.720	0.8858

9	n-10-10	6.8538	10.505	0.9597
10	n-10-15	8.2192	13.166	0.94805
11	n-10-20	7.2741	10.636	0.9563
12	n-10-25	6.8043	10.026	0.9615
13	n-5-15	7.2527	10.478	0.957
14	n-8-12	8.0416	11.838	0.9479
15	n-8-15	8.5387	13.629	0.9449
16	n-10-5-5	7.8119	10.884	0.9561
17	n-10-10-5	8.2065	11.957	0.9486
18	n-5-10-10	7.3472	10.571	0.9551
19	n-15-5-4	6.9293	11.266	0.9611
20	n-25-10-5	7.9419	12.446	0.9477

As shown in Table 2, network 12 (n-10-25) yielded the best performance among all the tested networks. Therefore this ANN model is recommended for predicting dew-point pressure of retrograde gas condensate reservoirs, when reservoir temperature, fluid composition, specific gravity of heptane plus, and molecular weight of heptane plus fraction are known. Fig. 3 shows the architecture of the recommended network.

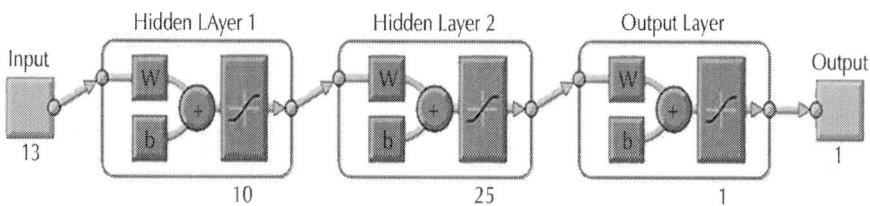

Figure 3: Proposed network topology for prediction of dew-point pressure.

RESULTS AND DISCUSSION

To evaluate the performance of the proposed model two types of analysis is performed. First the performance and accuracy of the new model in prediction of dew-point pressure is compared against existing literature correlations. Second, the validity and sensitivity of the proposed ANN model against variation in temperature is investigated.

Comparison of the ANN Model against Other Correlations

The data sets used to develop the ANN model were utilized to evaluate the performance and accuracy of the model against existing dew-point pressure correlations: Nemeth and Kennedy (1967), Elsharkawy (2002), and Shokir (2008). Both statistical and graphical means were used in this comparative evaluation. The correlation equations are presented in Appendix B.

The cross-plots of experimental versus predicted dew-point pressure are presented in Fig. 4, Fig. 5, Fig. 6 and Fig. 7. These cross-plots show the degree of agreement between experimentally measured data and the predicted values. As illustrated in Fig. 7, it evidences that the predictions of the dew-point pressures made by proposed ANN model yields the closest agreement with the experimental data among the selected correlations. Elsharkawy (2002) correlation shows largest number of scattered points in Fig. 3, which is reflected by the lowest correlation coefficient obtained ($R^2 = 0.76$).

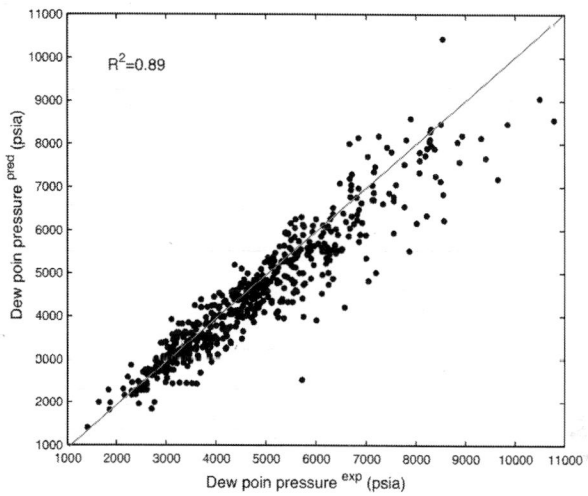

Figure 4: Cross plot for dew-point prediction (Nemeth and Kennedy, 1967).

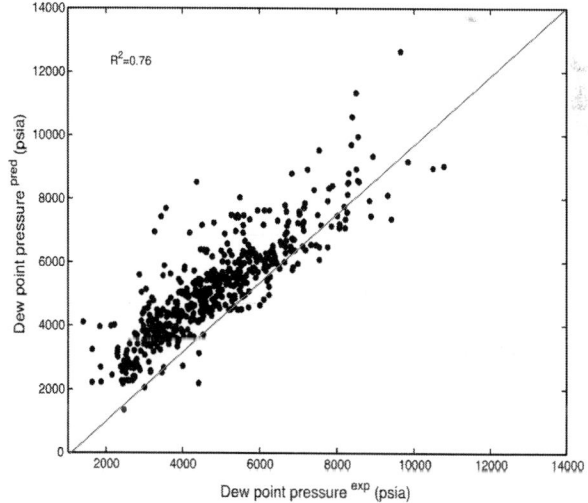

Figure 5: Cross plot for dew-point prediction (Elsharkawy, 2002).

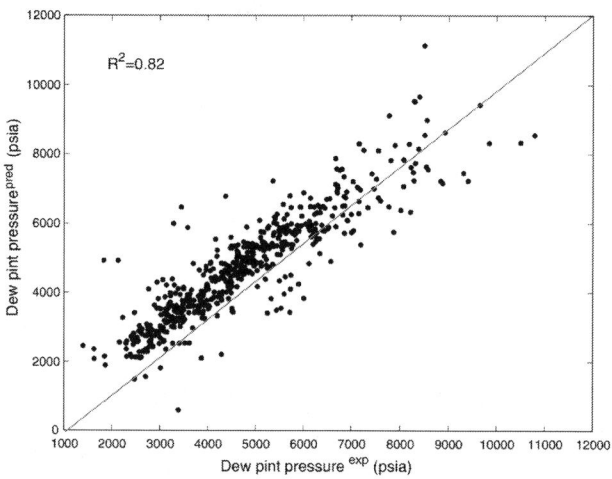

Figure 6: Cross plot for dew-point prediction (Shokir, 2008).

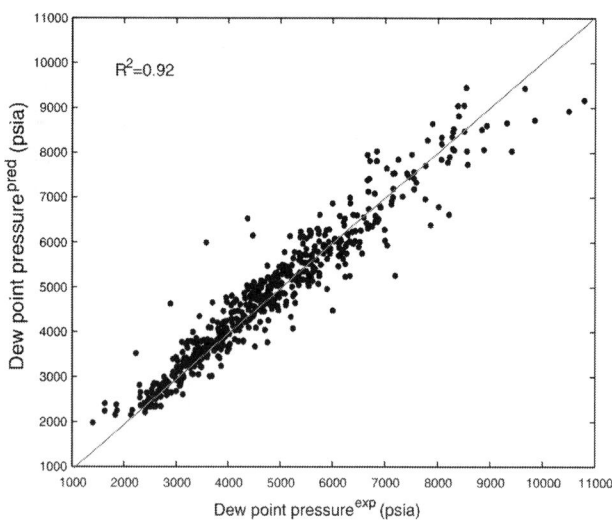

Figure 7: Cross plot for dew-point prediction (ANN (network 12)).

Error analysis shows that for ANN model the average absolute percent relative error (AAPE), root mean square error (RMSE), and R^2 values were 6.8043%, 10.026%, and 0.92%, respectively. While in the case of other correlations, higher values of AAPE, RMSE, and lower R^2 were obtained (Table 3).

Table 3: Statistical analysis results of ANN and correlations for dew-point pressure prediction

Model	APE	Max. APE	Min. APE	AAPE	RMSE	R^2
Nemeth and Kennedy (1967)	4.794	55.839	0.012	8.615	11.265	0.89
Elsharkawy (2002)	−9.667	192.367	0.030	15.363	23.773	0.76
Shokir (2008)	−0.728	168.692	0.022	11.019	17.791	0.82
ANN (this study)	−1.468	67.638	0.012	6.804	10.026	0.92

Table 4 lists 20 data sets from all 562 experimental data sets that were used in this study. Table 5 shows comparison of ANN results with Nemeth and Kennedy (1967), Elsharkawy (2002), and Shokir (2008)correlations for the data sets provided in Table 4. As can be seen, the proposed ANN model is capable of predicting the dew-point pressure with acceptable accuracy, while, Elsharkawy (2002) correlation shows large deviations from the experimental values. However, among the predictive correlations, predictions made by Nemeth and Kennedy (1967) correlation are nearly in good agreement with the experimental data.

Table 4: Some data sets of the experimental data used in this study

	T	C_1	C_2	C_3	C_4	C_5	C_6	C_{7+}	N_2	CO_2	H_2S	SG_{C7+}	MW_{C7+}	Pd
1	218	0.8686	0.0237	0.015	0.0133	0.0086	0.01	0.058	0	0.0028	0	0.8681	178	8540
2	240	0.8079	0.044	0.0288	0.203	0.014	0.0115	0.057	0.004	0.0125	0	0.7976	151	5720
3	265	0.8061	0.046	0.0238	0.019	0.0116	0.0088	0.07	0.0032	0.0115	0	0.824	194	6668
4	270	0.899	0.0542	0.0234	0.0133	0.0063	0.0053	0.0327	0.0035	0.0123	0	0.8324	200	10790
5	233	0.9668	0.0119	0.0032	0.0023	0.0028	0.0014	0.0057	0	0.0059	0	0.8597	224	9655
6	232	0.9359	0.0333	0.0094	0.0047	0.0022	0.0013	0.0056	0.0017	0.0059	0	0.8185	182	4372
7	129	0.9553	0.0155	0.0056	0.0026	0.013	0.0015	0.0019	0.0163	0	0	0.7617	126	1405
8	230	0.3344	0.05	0.0424	0.0322	0.0198	0.0109	0.0243	0.4322	0.0497	0.0041	0.768	128	4423
9	270	0.8634	0.0464	0.0183	0.0102	0.0031	0.0028	0.0331	0.002	0.0207	0	0.8368	235	8500
10	144	0.0349	0.0037	0.0018	0.0022	0.0023	0.002	0.0049	0.029	0.9192	0	0.8031	157	1835
11	250	0.7406	0.0772	0.0485	0.031	0.0175	0.0151	0.0606	0	0.0095	0	0.7805	130	3385
12	191	0.9646	0.0216	0.0037	0.0023	0.0006	0.0005	0.0046	0	0.0021	0	0.7821	175	3445
13	220	0.898	0.0381	0.0151	0.008	0.004	0.0045	0.0246	0	0.0077	0	0.7936	158	5870
14	130	0.9522	0.0168	0.0091	0.059	0.0027	0.0025	0.0098	0	0.001	0	0.7627	127	3520
15	194	0.6253	0.1288	0.0684	0.0332	0.0189	0.0132	0.1001	0.0035	0.0085	0.0001	0.8	161	4630
16	217	0.6795	0.056	0.0237	0.0208	0.0121	0.0147	0.0788	0.0088	0.0053	0.0942	0.7925	135	4414
17	176	0.5275	0.0348	0.0082	0.0055	0.0045	0.0045	0.0257	0.0104	0.0803	0.2986	0.836	140	3915
18	288	0.5685	0.1109	0.0858	0.0481	0.0221	0.0152	0.0912	0	0.0582	0	0.8312	157	4465
19	165	0.6627	0.0528	0.0216	0.0132	0.008	0.007	0.0304	0.0278	0.051	0.1255	0.7942	132	3564
20	160	0.9451	0.0168	0.0092	0.0061	0.0031	0.0032	0.0155	0	0.001	0	0.7627	127	3825

Table 5: Comparison of the ANN prediction results of the dew-point pressure with the other correlations

	Experimental dew-point	Nemeth and Kennedy (1967)		Elsharkawy (2002)		Shokir (2008)		ANN (This study)	
		Predicted dew-point	Relative error (%)	Predicted dew-point	Relative error (%)	Predicted dew-point	Relative error (%)	Predicted dew-point	Relative error (%)
1	8540	10,433.34	-22.17	8571.27	-0.37	7636.65	10.58	9446.74	-10.62
2	5720	2526.00	55.84	4568.18	20.14	3425.40	40.12	5789.50	-1.22
3	6668	8003.85	-20.03	7788.25	-16.80	7869.69	-18.02	7946.66	-19.18
4	10790	8556.35	20.70	9031.43	16.30	8542.15	20.83	9163.11	15.08
5	9655	7190.12	25.53	12646.63	-30.99	9424.22	2.39	9431.91	2.31
6	4372	5196.45	-18.86	8516.08	-94.79	6782.75	-55.14	6525.79	-49.26
7	1405	1405.43	-0.03	4107.76	-192.37	2452.84	-74.58	1979.01	-40.85
8	4423	3851.83	12.91	2193.54	50.41	4773.22	-7.92	4814.13	-8.84
9	8500	7133.63	16.07	11342.72	-33.44	11130.68	-30.95	9049.73	-6.47
10	1835	2285.81	-24.57	3964.19	-116.03	4930.50	-168.69	2149.29	-17.13
11	3385	3633.08	-7.33	3862.21	-14.10	4103.08	-21.21	3800.76	-12.28
12	3445	3295.67	4.33	4314.64	-25.24	3413.58	0.91	3424.39	0.60
13	5870	5728.09	2.42	6279.77	-6.98	5936.59	-1.13	5911.36	-0.70
14	3520	2436.30	30.79	3976.79	-12.98	2532.35	28.06	3498.72	0.60
15	4630	4060.53	12.30	4477.27	3.30	4572.19	1.25	4579.49	1.09
16	4414	4138.07	6.25	4146.24	6.07	4546.57	-3.00	4267.20	3.33

17	3915	4246.56	−8.47	4214.84	−7.66	3956.57	−1.06	3998.12	−2.12
18	4465	4101.60	8.14	4818.99	−7.93	4734.54	−6.04	4407.54	1.29
19	3564	3746.25	−5.11	3743.63	−5.04	3806.07	−6.79	3633.20	−1.94
20	3825	3489.65	8.77	4337.43	−13.40	3617.64	5.42	3822.83	0.06

Validity and Sensitivity of the Proposed ANN Model against Variation in Temperature

Four reservoir fluid samples, whose compositions are provided in Table 6, were used to check the validity and sensitivity of the proposed ANN model to reservoir temperature. Fig. 8 shows the effect of temperature on dew-point pressure of the samples. It is evident that predictions made by ANN model are in closest agreement with the experimental data. While in the case of other correlations, for example Elsharkawy (2002), predictions are largely separated from the experimental data points. On the other hand, Nemeth and Kennedy (1967) and Elsharkawy (2002) predictions show an increasing trend with increasing temperature for all of the samples studied. This is because during developing of these correlations an exponential (Eq. (B.1)) and linear relationship (Eq. (B.2)) between dew-point pressure and temperature were considered, respectively. On the contrary, Shokir (2008) correlation do not show specific trend with temperature. As a matter of fact, for retrograde gas condensate reservoir with a constant reservoir fluid composition, by increasing temperature, dew-point pressure increases to a certain point (i.e., cricondenbar). Beyond this point the dew-point pressure decreases with increasing temperature. It is evident that the proposed ANN model captures the physical trend of increasing dew-point pressure with the rise of temperature to certain point, and that after this point has been reached, then the dew point pressure decreases (see Fig. 8c and d). None of the studied correlations capture this trend. So this indicates that the model successfully simulates the actual physical process.

Table 6: Samples used for investigation of effect of temperature on dew-point pressure

Composition	Mole fraction			
	Sample A	Sample B	Sample C	Sample D

H$_2$S	0	0	0	0
CO$_2$	0.001	0.0079	0.0044	0
N$_2$	0.0417	0	0.0030	0.0058
C$_1$	0.8535	0.9141	0.8150	0.788
C$_2$	0.0477	0.0381	0.0372	0.059
C$_3$	0.0239	0.0147	0.0144	0.0315
C$_4$	0.0108	0.0072	0.1020	0.0266
C$_5$	0.0059	0.0031	0.0501	0.0425
C$_6$	0.0046	0.0029	0.0354	0.0252
C$_{7+}$	0.0109	0.012	0.0303	0.0214
MW$_{C7+}$	123	158	106	106
SG$_{C7+}$	0.7549	0.7936	0.733	0.733

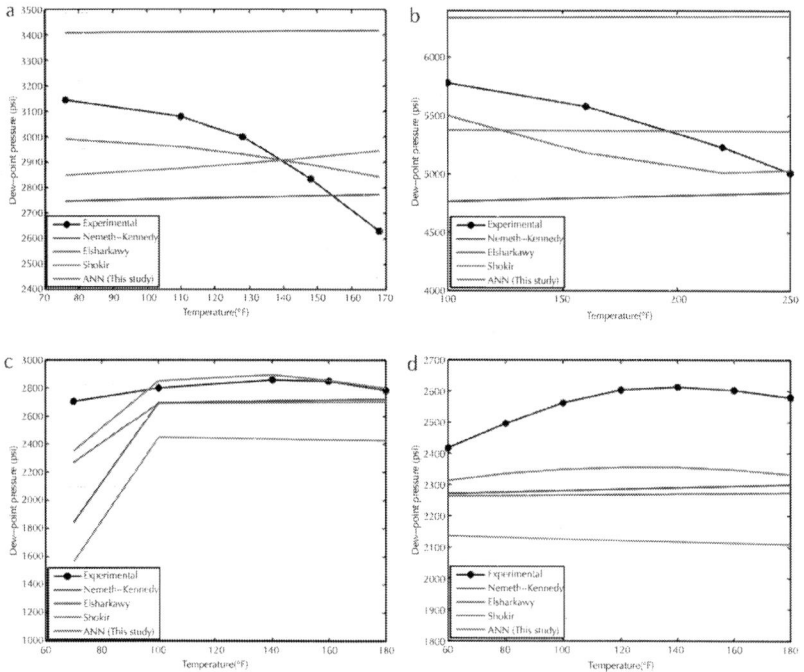

Figure 8: Effect of temperature on dew-point pressure: (a) sample 1, (b) sample 2, (c) sample 3, (d) sample 4.

Outlier Detection

Outlier diagnostics is an important step in developing the mathematical models (Goodall, 1993, Gramatica, 2007, Rousseeuw and Leroy, 2005 and Shokrollahi et al., 2013). Detection of outlier is to identify of groups of data that may differ from the bulk of the data present in a dataset. As a consequence, there is indeed a need to evaluate the available experimental data for dew-point pressure of gas condensate data. Since uncertainties affect the prediction capability of the developed model. To this end, we have applied the method of leverage value statistics (Goodall, 1993, Gramatica, 2007, Rousseeuw and Leroy, 2005 and Shokrollahi et al., 2013). The Graphical detection of the suspended (doubtful) data or outliers is undertaken through sketching the Williams plot on the basis of the calculated H values (Farasat et al., 2013, Goodall, 1993, Gramatica, 2007 and Rousseeuw and Leroy, 2005). A detailed description of equations and computational procedure of this method can be found elsewhere (Farasat et al., 2013, Goodall, 1993, Gramatica, 2007 and Rousseeuw and Leroy, 2005). The Williams plot has been sketched inFig. 9 for the results using the proposed ANN model. Existence of the majority of data points in the ranges $0 < H < 0.075$ and $-3 < Res < +3$ reveals that the applied models are statistically correct and valid. Furthermore, it shows that the whole data except eleven points in the dataset are located within the applicability domains of the applied models. Therefore, there are 11 points in the datasets which are within this the domain of $Res > +3$ and $Res < -3$ and consequently we can state it as probable doubtful datum.

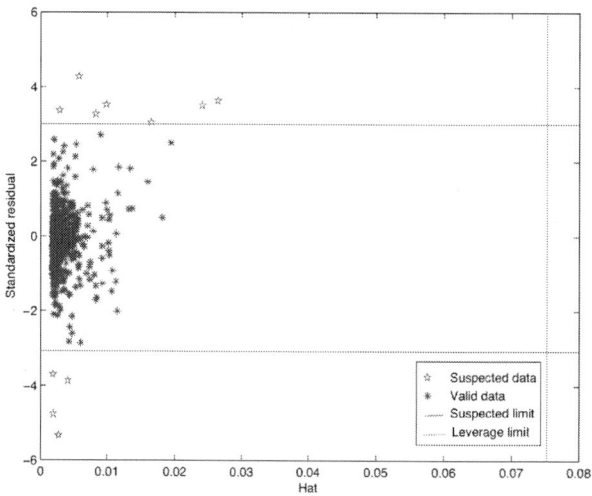

Figure 9: Detection of probable outliers and applicability domain of the presented ANN model.

CONCLUSIONS

Application of artificial neural network to predict the dew-point pressure of retrograde gas condensate reservoirs as a function of reservoir temperature, hydrocarbon and non-hydrocarbon reservoir fluid compositions, and characteristics of heptane plus fraction was studied in this work. The method utilized wide range of experimental data points to develop a more accurate predictive tool. After a series of optimization processes by monitoring the networks performance the best network structure was obtained. The final selected architecture was feed forward neural network with three hidden layers: first, second, and third layers having 10, 25, and 1 neurons, respectively. As shown by the statistical error analysis and graphical cross-plots, proposed neural network method developed yields better performance in comparison to the best available correlations. Error analysis showed that for ANN model the average absolute percent relative error (AAPE) of 6.8%, root mean square error (RMSE) of 10.0%, and correlation coefficient

(R^2) of 0.92. This improved ANN model is capable of capturing the physical trend of the dew-point pressure versus temperature for a constant composition fluid.

REFERENCES

1. Ahmadi, M.A., Ebadi, M., Shokrollahi, A., Majidi, S.M.J., 2013. Evolving artificial neural network and imperialist competitive algorithm for prediction oil flow rate of the reservoir. Appl. Soft Comput. 13, 1085–1098.

2. Ali Ahmadi, M., Zendehboudi, S., Lohi, A., Elkamel, A., Chatzis, I., 2013. Reservoir permeability prediction by neural networks combined with hybrid genetic algorithm and particle swarm optimization. Geophys. Prospect. 61, 582–598.

3. Arabloo, M., Shokrollahi, A., Gharagheizi, F., Mohammadi, A.H., 2013. Toward a predictive model for estimating dew point pressure in gas condensate systems. Fuel Process. Tech. 116, 317–324.

4. Grieves, R.B., Thodos, G., 1963. The cricondentherm and cricondenbar temperatures of multicomponent hydrocarbon mixtures. SPE Journal 3, 287–292.

5. Chamkalani, A., Nareh'ei, M.A., Chamkalani, R., Zargari, M.H., Dehestani-Ardakani, M.R., Farzam, M., 2013. Soft computing method for prediction of CO_2 corrosion in flow lines based on neural network approach. Chem. Eng. Commun. 200, 731–747.

6. Chouai, A., Laugier, S., Richon, D., 2002. Modeling of thermodynamic properties using neural networks: application to refrigerants. Fluid Phase Equilib. 199, 53–62.

7. Crogh, A., 1996. Improved Correlations for Retrograde Gases. Texas A & M University, College Station, TX.

8. Elsharkawy, A.M., 2002. Predicting the dew point pressure for gas condensate reservoirs: empirical models and equations of state. Fluid Phase Equilib. 193, 147–165.

9.   Farasat, A., Shokrollahi, A., Arabloo, M., Gharagheizi, F., Mohammadi, A.H., 2013. Toward an intelligent approach for determination of saturation pressure of crude oil. Fuel Process. Technol. 115, 201–204.

10. Goodall, C.R., 1993. Computation using the QR decomposition. Handbook of Statistics, vol. 9. Elsevier Science Publishers, Amsterdam, The Netherlands, pp. 467–508.

11. Gramatica, P., 2007. Principles of QSAR models validation: internal and external. QSAR Comb. Sci. 26, 694–701.

12. Haykin, S., 1994. Neural Networks: A Comprehensive Foundation. Prentice Hall PTR, Upper Saddle River, NJ.

13. Hemmati-Sarapardeh, A., Shokrollahi, A., Tatar, A., Gharagheizi, F., Mohammadi, A.H., Naseri, A., 2013. Reservoir oil viscosity determination using a rigorous approach. J. Fuel 116, 39–48.

14. Humoud, A., Al-Marhoun, M., 2001. A New Correlation for Gas-Condensate Dewpoint Pressure Prediction. SPE Middle East Oil Show, Bahrain.

15. Kumar, K.V., 2009. Neural network prediction of interfacial tension at crystal/solution interface. Ind. Eng. Chem. Res. 48, 4160–4164.

16. Kurata, F., Katz, D.L.V., 1942. Critical Properties of Volatile Hydrocarbon Mixtures. University of Michigan, Ann Arbor, MI. Levenberg, K., 1944. A method for the solution of certain problems in least squares. Q. Appl. Math. 2, 164–168.

17. Marquardt, D.W., 1963. An algorithm for least-squares estimation of nonlinear parameters. J. Soc. Ind. Appl. Math. 11, 431–441.

18. Martin, J.J., 1979. Cubic equations of state—which? Ind. Eng. Chem. Fundam. 18, 81–97.

19. Mohammadi, A.H., Richon, D., 2007. Use of artificial neural networks for estimating water content of natural gases. Ind. Eng. Chem. Res. 46, 1431–1438.

20. Moradi, G.R., Khoshmaram, A.A., Riazi, M.R., 2011. Estimation of properties distribution of C7+ by using artificial neural networks. J. Pet. Sci. Eng. 76, 57–62.

21. Morooka, C.K., Guilherme, I.R., Mendes, J.R.P., 2001. Development of intelligent systems for well drilling and petroleum production. J. Pet. Sci. Eng. 32, 191–199.

22. Nemeth, L., Kennedy, H., 1967. A correlation of dewpoint pressure with fluid composition and temperature. Old SPE J. 7, 99–104.

23. Olds, R., Sage, B., Lacey, W., 1945. Volumetric and phase behavior of oil and gas from Paloma field. Trans. AIME 160, 77–99.

24. Olds, R., Sage, B., Lacey, W., 1949. Volumetric and viscosity studies of oil and gas from a San Joaquin valley field. Trans. AIME 179, 287–302.

25. Organick, E.I., Golding, B.H., 1952. Prediction of saturation pressures for condensate-gas and volatile-oil mixtures. J. Pet. Tech. 4, 135–148.

26. Pedersen, K.S., Thomassen, P., Fredenslund, A., 1988. Characterization of Gas Condensate Mixtures.

27. Rafiee-Taghanaki, S., Arabloo, M., Chamkalani, A., Amani, M., Zargari, M.H., Adelzadeh, M.R., 2013. Implementation of SVM framework to estimate PVT properties of reservoir oil. Fluid Phase Equilib. 346, 25–32.

28. Reamer, H.H., Sage, B.H., 1950. Volumetric behavior of oil and gas from a Louisiana field I. J. Pet. Tech. 189, 261–268.

29. Rój, E., Wilk, M., 1998. Simulation of an absorption column performance using feed-forward neural networks in nitric acid production. Comput. Chem. Eng. 22 (Suppl 1), S909–S912.

30. Roosta, A., Setoodeh, P., Jahanmiri, A., 2011. Artificial neural network modeling of surface tension for pure organic compounds. Ind. Eng. Chem. Res. 51, 561–566.

31. Rousseeuw, P.J., Leroy, A.M., 2005. Robust Regression and Outlier Detection. Wiley, New York, NY.

32. Sage, B., Olds, R., 1947. Volumetric behavior of oil and gas from several San Joaquin Valley Fields. Trans. AIME 170, 156.

33. Sarkar, R., Danesh, A.S., Todd, A.C., 1991. Phase behavior modeling of gas-condensate fluids using an equation of state,

34. SPE Annual Technical Conference and Exhibition, Society of Petroleum Engineers, Inc., Dallas, TX.

35. Schalkoff, R.J., 1997. Artificial Neural Networks. McGraw-Hill Higher Education, New York, NY.

36. Shokir, E.M.E.-M., 2008. Dewpoint pressure model for gas condensate reservoirs based on genetic programming. Energy Fuels 22, 3194–3200.

37. Shokrollahi, A., Arabloo, M., Gharagheizi, F., Mohammadi, A.H., 2013. Intelligent model for prediction of CO2—reservoir oil minimum miscibility pressure. Fuel 112, 375–384.

38. Soave, G., 1972. Equilibrium constants from a modified Redlich–Kwong equation of state. Chem. Eng. Sci. 27, 1197–1203.

39. Turan, N.G., Mesci, B., Ozgonenel, O., 2011. The use of artificial neural networks (ANN) for modeling of adsorption of Cu(II) from industrial leachate by pumice. Chem. Eng. J. 171, 1091–1097.

40. Wilamowski, B., Iplikci, S., Kayank, O., Efe, M., 2001. International Joint Conference on Neural Networks (IJCNN '01), Washington, DC, July 15−19 , pp. 1778–1782, There is no corresponding record for this reference.

41. Zendehboudi, S., Ahmadi, M.A., James, L., Chatzis, I., 2012. Prediction of condensate-to-gas ratio for retrograde gas condensate reservoirs using artificial neural network with particle swarm optimization. Energy Fuels 26, 3432–3447.

42. Zudkevitch, D., Joffe, J., 1970. Correlation and prediction of vapor–liquid equilibria with the Redlich–Kwong equation of state. AIChE J. 16, 112–119.

Citations

CHAPTER 1

Laura R. Jarboe, Xueli Zhang, Xuan Wang, Jonathan C. Moore, K. T. Shanmugam, and Lonnie O. Ingram, "Metabolic Engineering for Production of Biorenewable Fuels and Chemicals: Contributions of Synthetic Biology," Journal of Biomedicine and Biotechnology, vol. 2010, Article ID 761042, 18 pages, 2010. doi:10.1155/2010/761042.

CHAPTER 2

Nianyin Li, Qian Zhang, Yongqing Wang, Pingli Liu, and Liqiang Zhao, "A New Multichelating Acid System for High-Temperature

Sandstone Reservoirs," Journal of Chemistry, Article ID 594913, in press.

CHAPTER 3

Isabel Natalia Sierra-Garcia and Valéria Maia de Oliveira (2013). Microbial Hydrocarbon Degradation: Efforts to Understand Biodegradation in Petroleum Reservoirs, Biodegradation - Engineering and Technology, Dr. Rolando Chamy (Ed.), ISBN: 978-953-51-1153-5, InTech, DOI: 10.5772/55920.

CHAPTER 4

Junjing Zhang, Liangchen Ouyang, D. Zhu, A.D. Hill, Experimental and numerical Studies of reduced fracture conductivity due to proppant embedment in the shale reservoir, Journal of Petroleum Science and Engineering, Available online 10 April 2015, ISSN 0920-4105, http://dx.doi.org/10.1016/j.petrol.2015.04.004.

CHAPTER 5

Mohammad Ali Ahmadi, Mohammad Ebadi, Arash Yazdanpanah, Robust intelligent tool for estimating dew point pressure in retrograded condensate gas reservoirs: Application of particle swarm optimization, Journal of Petroleum Science and Engineering, Volume 123, November 2014, Pages 7-19, ISSN 0920-4105, http://dx.doi.org/10.1016/j.petrol.2014.05.023.

CHAPTER 6

W.A. England, Reservoir geochemistry — A reservoir engineering perspective, Journal of Petroleum Science and Engineering, Volume

58, Issues 3–4, September 2007, Pages 344-354, ISSN 0920-4105, http://dx.doi.org/10.1016/j.petrol.2005.12.012.

CHAPTER 7

D.Y. Ding, Y.S. Wu, L. Jeannin, Efficient simulation of hydraulic fractured wells in unconventional reservoirs, Journal of Petroleum Science and Engineering, Volume 122, October 2014, Pages 631-642, ISSN 0920-4105, http://dx.doi.org/10.1016/j.petrol.2014.09.005.

CHAPTER 8

Naveed Arsalan, Jan J. Buiting, Quoc P. Nguyen, Surface energy and wetting behavior of reservoir rocks, Colloids and Surfaces A: Physicochemical and Engineering Aspects, Volume 467, 20 February 2015, Pages 107-112, ISSN 0927-7757, http://dx.doi.org/10.1016/j.colsurfa.2014.11.024.

CHAPTER 9

Seyed Mohammad Javad Majidi, Amin Shokrollahi, Milad Arabloo, Ramin Mahdikhani-Soleymanloo, Mohsen Masihi, Evolving an accurate model based on machine learning approach for prediction of dew-point pressure in gas condensate reservoirs, Chemical Engineering Research and Design, Volume 92, Issue 5, May 2014, Pages 891-902, ISSN 0263-8762, http://dx.doi.org/10.1016/j.cherd.2013.08.014.

Index